Order, Disorder and Criticality

Advanced Problems of Phase Transition Theory

Volume 4

Order, Disorder and Criticality

Advanced Problems of Phase Transition Theory

Volume 4

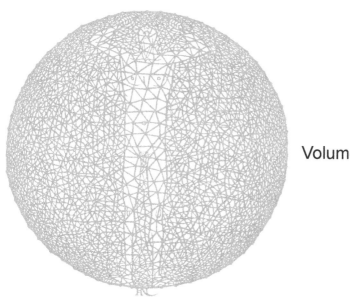

Editor

Yurij Holovatch

National Academy of Sciences, Ukraine

World Scientific

NEW JERSEY · LONDON · SINGAPORE · BEIJING · SHANGHAI · HONG KONG · TAIPEI · CHENNAI

Published by

World Scientific Publishing Co. Pte. Ltd.

5 Toh Tuck Link, Singapore 596224

USA office: 27 Warren Street, Suite 401-402, Hackensack, NJ 07601

UK office: 57 Shelton Street, Covent Garden, London WC2H 9HE

British Library Cataloguing-in-Publication Data
A catalogue record for this book is available from the British Library.

The design of the cover incorporates the artwork of Jacques Hnizdovsky (Constructor, Woodcut, 1967).

ORDER, DISORDER AND CRITICALITY
Advanced Problems of Phase Transition Theory
Volume 4

ISBN 978-981-4632-67-6

In-house Editor: Christopher Teo

Printed in Singapore

Preface

This volume continues a review series on phase transitions and critical phenomena that have been published by the World Scientific since 2004.[1] The principal goal of the series is to provide pedagogical introductory reviews of problems that are under current discussion in the scientific literature and are related to phase transitions and criticality.

Having appeared in the form of different variants of a mean field theory for the description of critical points in liquids and equilibrium phase transitions in connection with problems of statistical physics and thermodynamics, the theory of critical phenomena now goes far beyond its applications in pure physics. Physics is an archetype of a natural science with its primacy of experiment, intrinsic experiment-theory-simulation interplay, the use of mathematics as a tool and a language of study, and, last but not least, the modelling of natural phenomena. The above factors have led to the phenomenon we have been witnessing for the past several decades: physics (or at least physicists) have penetrated into a much wider field, analysing problems that were, originally, traditional subjects of different sciences. The same holds true for the theory of phase transitions and critical phenomena: the models used for their description and the methods applied allow for a conceptual understanding and interpretation of a wide range of phenomena in various systems, not necessarily of physical origin. This especially concerns complex systems, where the new, emergent properties appear via collective behaviour of simple elements. Since all complex systems involve cooperative behaviour between many interconnected components, the field of phase transitions and critical phenomena provides a very natural conceptual and methodological framework for their study.

The book starts with two chapters where genuine physical problems are treated by analytical and numerical approaches. In the first chapter,

---

[1] *Order, Disorder and Criticality. Advanced Problems of Phase Transition Theory*, edited by Yu. Holovatch (World Scientific, Singapore), vol. 1 – 2004; vol. 2 – 2007, vol. 3 – 2012.

Bertrand Berche and Ralph Kenna report on new developments concerning scaling above the upper critical dimension. They reconsider the question of hyperscaling and provide convincing arguments in favour of the introduction of a new scaling relation analogous to one introduced by Michael Fisher 50 years ago. Accordingly, this includes a new universal critical exponent (denoted ϙ) that governs scaling of the correlation length above the upper critical dimension. The new hyperscaling relation introduced by the authors is valid in all dimensions. The authors show how the new paradigm is supported by the renormalization group formalism. Whereas the problem considered in the first chapter is of fundamental importance for our understanding of critical phenomena in general, the second chapter of the book, written by Oleg Vasilyev, deals with the particular effect that arises in the vicinity of the critical point - an emergence of critical Casimir forces. Originally, the attractive Casimir force was predicted to act between two perfectly conducting plates in a vacuum. Its physical origin is the suppression of zero-level quantum electromagnetic fluctuations. However, it is now well established that such quantum effects have their classical counterpart in condensed matter physics. There, in the vicinity of the second-order phase transition, the long-range fluctuations of the order parameter lead to the so called critical Casimir effect. The chapter provides a description of numerical methods for the computation of critical Casimir interactions for lattice models in different geometries.

A common feature of the next two chapters of the book, written by Ralf Metzler and by Héléna Zapolsky, is that phenomena treated there, although being primarily analyzed in physical applications, are quite general and occur in various natural systems ranging from the physical to the biological. These are diffusion and pattern formation. Another common aspect that places them in the context of a review on advanced problems of phase transitions is that scaling, universality, and emergent behaviour are inherent properties allowing for their interpretation. The concept of diffusion is frequently used to provide accurate qualitative descriptions of various phenomena where statistics of random walks is essential. A chapter by Ralf Metzler provides an overview in the field of anomalous diffusion, where classical Brownian motion scaling is violated. Moreover, anomalous diffusion processes are not governed by the central limit theorem, which leads, in turn, to their unusual properties. In particular, the author concentrates on non-ergodic and ageing behaviour. Diffusion plays an important role in the phenomenon of pattern formation, often being one of the ingredients that leads to self-organization and that is behind similar patterns in nature. Mechanisms of pattern formation are treated in the chapter by Héléna Zapolsky. The chapter provides a brief historical overview of experimental observations of pattern formation and their theoretical description. Particular emphasis has been placed on computational modelling for the description of pattern formation dynamics. Special attention is paid to the stability of patterns with stripe and honeycomb morphology.

The final chapter serves as a complement to the foregoing material and identifies another field in which the theory of phase transitions and critical behaviour is being applied - i.e. the application of physical concepts and methods for modelling of social behaviour (in other words, sociophysics). Serge Galam presents a model of voting in bottom-up hierarchical systems, using analytic and numerical approaches to describe an outcome of the voting procedure. To this end, an analogy with the real-space renormalization procedure is exploited. As far as social behaviour is concerned, the author does not always explicitly use physical terminology. However, and implicitly, the concepts of the renormalization group flow, the fixed point and its stability are present, and this approach appears very promising, adding one more link between criticality and universality in physical and social systems.

The review series *Order, Disorder and Criticality. Advanced Problems of Phase Transition Theory* originated from the lectures presented during the international workshop "Ising Lectures",[2] an event that occurs annually in Lviv, Ukraine. I am using this occasion to extend my cordial thanks to my colleagues, the authors of this volume, for lecturing at the workshop and for writing these reviews. Special thanks are in order to World Scientific for their interest in getting this book published and all the very helpful assistance provided during the preparation of the manuscript.

Yurij Holovatch
Institute for Condensed Matter Physics,
National Academy of Sciences of Ukraine,
Lviv, 19.12.2014

---

[2]See: http://www.icmp.lviv.ua/ising/

Contents of vol. 1

Contents of vol. 2

Contents of vol. 3

Contents

Chapter 1

Scaling and Finite-Size Scaling above the Upper Critical Dimension

Ralph Kenna*

*Applied Mathematics Research Centre, Coventry University,
Coventry CV1 5FB, England*

Bertrand Berche†

*Statistical Physics Group, Institut Jean Lamour, UMR CNRS 7198,
Université de Lorraine B.P. 70239, 54506 Vandœuvre lès Nancy Cedex,
France*

In the 1960's, four famous scaling relations were developed which relate
the six standard critical exponents describing continuous phase transitions in the thermodynamic limit of statistical physics models. They
are well understood at a fundamental level through the renormalization
group. They have been verified in multitudes of theoretical, computational and experimental studies and are firmly established and profoundly important for our understanding of critical phenomena. One of the
scaling relations, hyperscaling, fails above the upper critical dimension. There, critical phenomena are governed by Gaussian fixed points
in the renormalization-group formalism. Dangerous irrelevant variables
are required to deliver the mean-field and Landau values of the critical exponents, which are deemed valid by the Ginzburg criterion. Also
above the upper critical dimension, the standard picture is that, unlike
for low-dimensional systems, finite-size scaling is non-universal, at least
at the critical point. Here we report on new developments which indicate that the current paradigm is flawed and incomplete. In particular,
the introduction of a new exponent characterising the finite-size correlation length allows one to extend hyperscaling beyond the upper critical
dimension. Moreover, finite-size scaling is shown to be universal provided the correct scaling window is chosen. These recent developments
above the upper critical dimension also lead to the introduction of a new
scaling relation analogous to one introduced by Fisher 50 years ago and

---

*r.kenna@coventry.ac.uk
†bertrand.berche@univ-lorraine.fr

deliver a statistical physics explanation for the emergence of effective four dimensionality as characteristic of generic field theories.

Contents

1. Introduction

In the standard picture, critical phenomena above the upper critical dimension are believed to be well described by mean-field theory and by Landau theory. This picture is supported by the renormalization-group formalism coupled to Fisher's dangerous irrelevant variables concept. In the conventional paradigm, hyperscaling fails above the upper critical dimension, which is four for many field theories, and finite-size scaling ceases to be universal at the critical point. In many reviews and textbooks, theories with short-range interactions beyond the upper critical dimension play a precursory role, appearing in the early, introductory, chapters, where they are seen as a step en route to more physically interesting phenomena at or below upper dimensionality. For this reason, high-dimensional theories play a foundational role in statistical mechanics. Increasing the interaction range reduces the upper critical dimension. Such experimentally accessible models may be of direct physical relevance. Over the last decades, however, it has been recognised that the current paradigm is not satisfactory. We quote from Ref. [1] in the context of ϕ^4 theory:

> "Thus we arrive at a rather disappointing state of affairs - although for the ϕ^4 theory in $d = 5$ dimensions all exponents are known, including those of the corrections to scaling, and in principle very complete analytical calculations are possible, the existing theories clearly are not so good."

Very recently, progress has been made to provide a complete account of scaling above the upper critical dimension. Given the basic role mean-field and Landau theories play, this development is important in a fundamental sense in various areas of physics. The aims of this exposition are to contextualise and present the recent theoretical advances. We begin in Section 2 with a summary of scaling and finite-size scaling at continuous phase transitions, including of the scaling relations. Derived in the 1960's, these form a cornerstone for the statistical physics of phase transitions. We also touch upon logarithmic corrections to scaling, a subject reviewed in a previous volume of this series.[2] In Section 3 we revisit Widom's scaling ansatz and Kadanoff's block-spin renormalization. Mean-field theory and Landau theory are developed as a foundation for ϕ^4 theory in Section 4, where the Ginzburg criterion is recalled. In Section 5, Wilson's renormalization-group formalism for the Gaussian fixed point is briefly outlined. To describe scaling above the upper critical dimension, dangerous irrelevant variables have to be accounted for, and these are treated in both infinite and finite systems in Section 6. Reasons why this standard paradigm is ad hoc and unsatisfactory are given in Section 7. Recent developments are summarised in Section 8, where two new exponents ϙ (pronounced "koppa") and η_Q are introduced. The new, full scaling and finite-size scaling theory above the upper critical dimension is outlined in Section 9. We also include new results for logarithmic corrections *at* the critical dimension itself, supplementing the review in Ref. [2]. We conclude in Section 10 with some implications for the special role played by four dimensionality in physics.

2. Scaling at Continuous Phase Transitions

In this section, we briefly summarise the standard power-law scaling paradigm at continuous phase transitions and the relations between the various critical exponents. This includes a description of standard finite-size scaling (FSS). We also very briefly summarise the new developments which are elucidated in subsequent sections.

Although second-order or continuous phase transitions also appear in particle physics, cosmology, fluid mechanics, and other areas of physics, we employ the language of magnetism in this exposition for purposes of clarity. It is straightforward to convert the results to the terminology of related fields.

To begin this section, we introduce the basic functions which describe global and local properties of the system. These are called thermodynamic and correlation functions respectively. Since the latter concern system details at a microscopic level, their definitions are model specific and throughout this work, we use the Ising and ϕ^4 models as generic representations of spin models and field theories.

2.1. *Thermodynamic functions*

We consider a system of spins s_i located at the sites i of a d-dimensional lattice with L sites in each direction, so that there are $N = L^d$ sites overall. If the lattice constant is a, the volume of the system is Na^d. The partition function is defined as

$$Z_L = \sum_{\{s_i\}} e^{-\beta \mathcal{H}}, \qquad (2.1)$$

where $\mathcal{H}$ is the total energy associated with a given configuration $\{s_i\}$ of the spins and $\beta = 1/k_B T$ is the reciprocal of the Boltzmann constant k_B times the absolute temperature T. The Hamiltonian is

$$\mathcal{H} = E - HM, \qquad (2.2)$$

in which E represents the energy due to interactions between the spins themselves, H is the strength of an external magnetic field and M is the magnetisation of a given configuration. For later convenience we define the reduced external field as $h = \beta H$.

The Ising model is defined through the configurational energy and magnetisation[3,4]

$$E = -J \sum_{\langle ij \rangle} s_i s_j, \tag{2.3}$$

$$M = \sum_i s_i, \tag{2.4}$$

in which $s_i \in \{\pm 1\}$. The Helmholtz free energy is defined as

$$F_L = -k_B T \ln Z_L. \tag{2.5}$$

We denote intensive quantities in lower case, so that, e.g.,

$$f_L = \frac{F_L}{N}. \tag{2.6}$$

At $h = 0$, the entropy is given by

$$N s_L = -\frac{\partial F_L}{\partial T} = k_B \ln Z_L + \frac{1}{T} N e_L, \tag{2.7}$$

where

$$N e_L = -\frac{\partial \ln Z_L}{\partial \beta} = \langle E \rangle, \tag{2.8}$$

is the internal energy and $\langle \ldots \rangle$ refer to expectation values. The specific heat is

$$c = \frac{\partial e}{\partial T} = \frac{k \beta^2}{N} \left(\langle E^2 \rangle - \langle E \rangle^2 \right). \tag{2.9}$$

The magnetisation and magnetic susceptibility are respectively defined as

$$m = -\frac{\partial f}{\partial H} = \frac{\langle M \rangle}{N}, \tag{2.10}$$

$$\chi = \frac{\partial m}{\partial H} = \frac{\beta}{N} \left(\langle M^2 \rangle - \langle M \rangle^2 \right). \tag{2.11}$$

2.2. *Correlation functions*

For the Ising model, the magnetisation defined in Eq.(2.10) is

$$m = \frac{1}{N} \sum_i \langle s_i \rangle. \tag{2.12}$$

In the translation-invariant case where $\langle s_i \rangle$ is independent of i, this simplifies to $m = \langle s_i \rangle$. This may be expected when the system is infinite in extent. There is no spontaneous magnetisation in a finite-size system. For the spin model, the connected correlation function is defined as

$$G(r_i, r_j) = \langle s_i s_j \rangle - \langle s_i \rangle \langle s_j \rangle, \tag{2.13}$$

where r_i represents the position of the ith lattice site. It is useful to promote h to a function of position, so that the external magnetic field is not uniform. One can easily show that the connected correlation function is

$$G(r_i, r_j) = -\frac{\beta}{N} \frac{\partial^2 f}{\partial h_i \partial h_j}, \tag{2.14}$$

where h_i is the reduced field at site i.

2.3. Power-law critical-point scaling

We are interested in behaviour near the critical point $(T, H) = (T_c, 0)$. To gain a dimensionless measure of distance from criticality we define the reduced temperature

$$t = \frac{T - T_c}{T_c}. \tag{2.15}$$

We henceforth write thermodynamic functions in terms of reduced variables, e.g., $c_L(t, h)$, $m_L(t, h)$ and $\chi_L(t, h)$. To make the connection to the macroscopic thermodynamical world, statistical mechanics is taken to its large L limit. There the leading power-law scaling behaviour which captures the dependencies of these thermodynamic functions at a phase transition of second order is given by

$$c_\infty(t, 0) \sim |t|^{-\alpha}, \tag{2.16}$$

$$m_\infty(t, 0) \sim (-t)^\beta, \tag{2.17}$$

$$m_\infty(0, h) \sim h^{\frac{1}{\delta}}, \tag{2.18}$$

$$\chi_\infty(t, 0) \sim |t|^{-\gamma}, \tag{2.19}$$

which define the critical exponents α, β, γ and δ. Eq.(2.17) holds only when $T < T_c$ as there is no spontaneous magnetisation when $T > T_c$.

The thermodynamic functions listed above are derivable from the partition function through Eqs.(2.8)-(2.11). They give the global response of the *entire* system to tuning the temperature and/or external field near the phase transition. One is also interested in local responses given by the correlation functions (2.13). For systems with translational invariance one expects $G(r_i, r_j)$ to depend only on the distance between sites i and j rather than their absolute coordinates. We write

$$G_\infty(t, h; r) \sim r^{-p} D\left[\frac{r}{\xi_\infty(t, h)}\right], \tag{2.20}$$

in which ξ represents the correlation length, which measures the scale of fluctuations away from the fully aligned or fully random states. The correlation length also diverges close to criticality, and

$$\xi_\infty(t, 0) \sim |t|^{-\nu}, \qquad \xi_\infty(0, h) \sim |h|^{-\nu_c}. \tag{2.21}$$

When $r \ll \xi_\infty(t, 0)$, the power-law dominates Eq.(2.20) and one writes

$$G_\infty(t, 0; r) \sim r^{-(d-2+\eta)}, \tag{2.22}$$

in which η is the anomalous dimension.

From experiments and Landau theory (Section 4) it is expected that the correlation function decays exponentially away from criticality for sufficiently large distance. Then $D(y) \sim \exp(-y)$, or

$$G_\infty(t, 0; r) \sim r^{-p} \exp\left[-\frac{r}{\xi_\infty(t, 0)}\right], \tag{2.23}$$

when $r \gg \xi_\infty(t, 0)$.

The momentum-space equivalent of the general form (2.20) is

$$\tilde{G}(t, h; k) = \int_0^\infty d^d \mathbf{r} \, G(r) e^{ikr} = k^{p-d} g(k\xi), \tag{2.24}$$

for some function g. In the critical region, where $p = d - 2 + \eta$, this becomes

$$\tilde{G}(t, 0; k) \sim k^{-2+\eta}. \tag{2.25}$$

2.4. *Fundamental theory of phase transitions*

In the 1950's, inspired by the fundamental theorem of algebra, Yang and Lee[5] developed a *fundamental theory of phase transitions*. For the finite L Ising model, for example, the partition function in Eq.(2.1) has a discrete set of zeros in the complex-h plane. In the limit of infinite volume these Lee-Yang zeros condense onto curves. In fact, in many circumstances the Lee-Yang theorem ensures that these zeros are purely imaginary.[5]

When $T > T_c$, the point in the distribution of Lee-Yang zeros closest to the real axis is called the Yang-Lee edge and we denote it by $h_{\text{YL}}(t)$. It approaches the real axis as t reduces to its critical value $t = 0$ and that approach is also characterised by a power law if the transition is second order, namely

$$h_{YL}(t) \sim t^{\Delta}, \tag{2.26}$$

in which Δ is called the gap exponent.

Fisher pointed out that the finite-size partition function in Eq.(2.1) also has zeros in the plane of complex temperature. In the thermodynamic limit these Fisher zeros[6] pinch the real temperature axis at the point where the phase transition occurs (namely at $T = T_c$).

2.5. *Finite-size scaling, shifting and rounding*

Genuine phase transitions can only appear in systems of infinite size. In finite-size systems, the divergences which are characteristic of some functions are replaced by peaks of finite height. These peaks have a "rounded" structure and are shifted away from the critical point of the infinite-volume system to what are called the pseudocritical points or effective critical points.

The FSS hypothesis is that the relationship between functions in the thermodynamic limit and their finite-size counterparts enters through the ratio of the two length scales L and ξ_{∞}.[8–11] For a generic function $P(t, h)$, say, we express this relationship as (setting $h = 0$)

$$\frac{P_L(t)}{P_{\infty}(t)} = \mathcal{F}_P \left[\frac{L}{\xi_{\infty}(t)} \right]. \tag{2.27}$$

Here, and henceforth, we have suppressed the dependency of P on h when the latter vanishes.

We suppose $P_{\infty}(t) \sim |t|^{-\rho}$ near the critical point in the thermodynamic limit. Fixing the scaling ratio $w = L/\xi_{\infty}(t)$ in Eq.(2.27) amounts to the substitution $|t| \to w^{1/\nu} L^{-1/\nu}$, from which

$$P_L(t) = \mathcal{F}_P(w) P_{\infty} \left(w^{\frac{1}{\nu}} L^{-\frac{1}{\nu}} \right) \sim L^{\frac{\rho}{\nu}}. \tag{2.28}$$

In fact the scaling form $P_L(t) \sim L^{\rho/\nu}$ usually holds as a good approximation in a region known as the *scaling window*. In many instances the scaling window includes *both* the critical point $t = 0$ as well as the pseudocritical point $t = t_L$, defined as the value of t at which $P_L(t)$ has an extremum. In other words, to obtain the FSS behaviour of P_L from $P_{\infty}(t)$, one simply replaces the infinite-volume correlation length $\xi_{\infty}(t)$ by the actual length of the system L inside the scaling window. Applying it to the functions in Eqs.(2.16), (2.17), (2.19), (2.21), (2.23), (2.26), one obtains

$$c_L(t_L) \sim L^{\frac{\alpha}{\nu}}, \tag{2.29}$$

$$m_L(t_L) \sim L^{\frac{-\beta}{\nu}}, \tag{2.30}$$

$$\chi_L(t_L) \sim L^{\frac{\gamma}{\nu}}, \tag{2.31}$$

$$\xi_L(t_L) \sim L, \tag{2.32}$$

$$G_L(t_L; r) \sim r^{-p} \exp\left(-\frac{r}{N_0 L}\right), \tag{2.33}$$

$$h_1(t) \sim L^{\frac{\Delta}{\nu}}. \tag{2.34}$$

In Eq.(2.33), N_0 represents an appropriate amplitude and we have omitted the subscript L from the finite-size scaling of the first Lee-Yang zero.

One is also interested in how the location of the peak $t = t_L$ is shifted relative to its infinite-volume limit $t = 0$ (the critical point). This is characterised by the so-called shift exponent λ_T. For a system of linear extent L the scaling of the pseudocritical point also follows a power-law

$$t_L \sim L^{\lambda_T}, \tag{2.35}$$

to leading order.

The smoothening out of the divergence present in the thermodynamic limit into a peak is also associated with the rounding exponent θ_T. The rounding may be defined as the width of the susceptibility curve at half of its maximum height, ΔT. One then has

$$\Delta T \sim L^{-\theta_T}, \tag{2.36}$$

to leading order.

2.6. *Scaling relations*

The six core critical exponents are related by four famous scaling relations, derived in the 1960's:

$$\nu d = 2 - \alpha, \tag{2.37}$$

$$2\beta + \gamma = 2 - \alpha, \tag{2.38}$$

$$\beta(\delta - 1) = \gamma, \tag{2.39}$$

$$\nu(2 - \eta) = \gamma. \tag{2.40}$$

The relation (2.37) was developed by Widom[12,14] using considerations of dimensionality, with alternative arguments given by Kadanoff.[15] Josephson[16] later derived the inequality $\nu d \geq 2-\alpha$ on the basis of some plausible thermo-dynamic assumptions. Because it involves dimensionality, Eq.(2.37) is also called the *hyperscaling relation*. It is conspicuous in the set (2.37)–(2.41) in that it is the only scaling relation involving d. The hyperscaling relation (2.37) lies at the heart of this review. The equality (2.38) was originally proposed by Essam and Fisher[17] and the related inequality $2\beta + \gamma \geq 2 - \alpha$ was rigorously proven by Rushbrooke.[18] Similarly, relation (2.39) was put forward by Widom[19] and the related inequality $\beta(\delta-1) \leq \gamma$ proved by Griffiths.[20] Equalities (2.38) and (2.39) were re-derived by Abe[21] and Suzuki[22] using an alternative route involving Lee-Yang zeros. Eq.(2.40) was derived by Fisher,[23] with a related inequality proved in Ref. [24,25]. This equation is also a focus of this exposition. The gap exponent is given by[26]

$$\Delta = \frac{\delta\gamma}{\delta - 1} = \delta\beta = \beta + \gamma. \tag{2.41}$$

The critical exponent governing the behaviour of the correlation length in field is

$$\nu_c = \frac{\nu}{\Delta}. \tag{2.42}$$

The prediction for the shifting exponent coming from standard FSS is

$$\lambda_T = \frac{1}{\nu}. \tag{2.43}$$

But this is not always true and in some cases, such as in the Ising model in two dimensions with special boundary conditions, it can deviate from this value. Similarly the rounding exponent is usually

$$\theta_T = \frac{1}{\nu}. \tag{2.44}$$

2.7. *Logarithmic corrections*

In certain circumstances, there are multiplicative logarithmic corrections to the leading behaviour and[2]

$$c_\infty(t) \sim |t|^{-\alpha} |\ln|t||^{\hat{\alpha}}, \tag{2.45}$$

$$m_\infty(t) \sim (-t)^{\beta} |\ln(-t)|^{\hat{\beta}} \quad \text{for } t < 0, \tag{2.46}$$

$$m_\infty(h) \sim h^{\frac{1}{\delta}} |\ln h|^{\hat{\delta}}, \tag{2.47}$$

$$\chi_\infty(t) \sim |t|^{-\gamma} |\ln|t||^{\hat{\gamma}}, \tag{2.48}$$

$$\xi_\infty(t) \sim |t|^{-\nu} |\ln|t||^{\hat{\rho}}, \tag{2.49}$$

$$r_{\text{YL}}(t) \sim t^{\Delta} |\ln t|^{\hat{\Delta}} \quad \text{for } t > 0. \tag{2.50}$$

In addition the scaling of the correlation function at $h = 0$ has

$$\mathcal{G}_\infty(r, t) \sim r^{-(d-2+\eta)}(\ln r)^{\hat{\eta}} D\left[\frac{r}{\xi_\infty(t)}\right]. \qquad (2.51)$$

To allow for logarithmic corrections in the correlation length of the finite-size system, we write

$$\xi_L(0) \sim L(\ln L)^{\hat{\mathbf{q}}}. \qquad (2.52)$$

Recently scaling relations between these logarithmic-correction exponents have been established.[27] Analogues of Eqs.(2.37)–(2.40) are (for a review see the previous volume in this series[2])

$$\hat{\alpha} = \begin{cases} 1 + d(\hat{\mathbf{q}} - \hat{\nu}) & \text{if} \quad \alpha = 0 \quad \text{and} \quad \phi \neq \pi/4 \\ d(\hat{\mathbf{q}} - \hat{\nu}) & \text{otherwise,} \end{cases} \qquad (2.53)$$

$$2\hat{\beta} - \hat{\gamma} = d(\hat{\mathbf{q}} - \hat{\nu}), \qquad (2.54)$$

$$\hat{\beta}(\delta - 1) = \delta\hat{\delta} - \hat{\gamma}, \qquad (2.55)$$

$$\hat{\eta} = \hat{\gamma} - \hat{\nu}(2 - \eta). \qquad (2.56)$$

In the first of these, ϕ refers to the angle at which the Fisher zeros impact onto the real axis. If $\alpha = 0$, and if this impact angle is any value other than $\pi/4$, an extra logarithm arises in the specific heat. For example, this happens in $d = 2$ dimensions, but not in $d = 4$, where $\phi = \pi/4$.[27] One notes the crucial role played by the exponent $\hat{\mathbf{q}}$ in the scaling relations for logarithmic corrections. The question arises, what is the analogue of this for the leading exponents, and why does it not appear in the usual scaling relations (2.37)–(2.40). That is the subject of much of what follows and next we summarise the answer.

2.8. *Q-Scaling and Q-FSS: The new paradigm*

For clarity and convenience, we gather here the new results recently derived for scaling and FSS above the upper critical dimension and reviewed in this Chapter.

Of core importance is the leading power-law analogue of $\hat{\mathbf{q}}$ in the correlation length,

$$\xi_L(t_L) \sim L^{\mathbf{q}}. \qquad (2.57)$$

We shall show that the new exponent $\mathbf{q}$ is given by

$$\mathbf{q} \sim \begin{cases} \frac{d}{d_c} & \text{if } d > d_c \\ 1 & \text{if } d < d_c \end{cases}. \qquad (2.58)$$

We will also establish the leading power-law analogue to Eq.(2.53) as

$$\frac{\nu d}{\varsigma} = 2 - \alpha. \tag{2.59}$$

This scaling relation holds in all dimensions and is the extension of hyper-scaling to $d > d_c$.

We will also show that two separate correlation functions are required above d_c. Which one to use depends upon whether one uses the scale of the lattice extent L or the correlation length ξ_L. In the former case one has at criticality

$$G_L(r) \sim r^{-d-2+\eta_Q}, \tag{2.60}$$

while in the latter case,

$$G_\xi(r) \sim r^{-d_c-2+\eta}. \tag{2.61}$$

The formula (2.22) is therefore only valid below d_c. The second new exponent η_Q is the anomalous dimension when distance is measured in the scale of L. It is related to η, the anomalous dimension on the scale of ξ, by

$$\eta_Q = \varsigma\eta + 2(1 - \varsigma). \tag{2.62}$$

Thus η_Q and η coincide when $d < d_c$. These anomalous dimensions also have logarithmic counterparts for the case when $d = d_c$, thus providing an additional formula to the list (2.53)–(2.56) and an amendment to Chapter 1 of Ref. [2]. (See Section [9.2] herein.)

In the following, we provide evidence that the new exponents are both *physical* and *universal*. They are physical in the sense that they control the finite-size behaviour of the correlation length and correlation function. We also suggest that they are universal, independent of the boundary conditions used for finite-size systems.

3. Widom Scaling and Kadanoff Renormalization as Bases for the Scaling Relations

Historically, scaling theory begins with Widom's scaling ansatz for the magnetisation[12] (see also Ref. [13])

$$m_\infty(t, h) = |t|^\beta \mathcal{M}_\pm\left(\frac{h}{|t|^\Delta}\right), \quad \text{for} \quad t \to 0^\pm, h \to 0. \tag{3.1}$$

Setting $|t| = h^{1/\Delta}$ allows one to re-express this as $m_\infty(t, h) \sim h^{\beta/\Delta}$. Comparing with Eq.(2.17), one identifies the gap exponent as

$$\Delta = \beta\delta. \tag{3.2}$$

To achieve the form (3.1), Widom suggested that the singular part of the free energy scale as

$$f_\infty(t, h) = |t|^{2-\alpha} \mathcal{F}_\pm \left(\frac{h}{|t|^\Delta} \right).$$ (3.3)

(This functional form of the scaling function ensures that h enters the partition function through the ratio $h/|t|^\Delta$. This means that the Lee-Yang zeros scale as $|t|^\Delta$, so that Δ is the gap exponent of Eq.(2.26).) A similar scaling ansatz can be written for the correlation function:[15]

$$G_\infty(r, t, h) \sim \frac{1}{r^{d-2+\eta}} \mathcal{G}_\pm \left(\frac{r}{\xi}, \frac{h}{|t|^\Delta} \right), \quad \text{for} \quad t \to 0^\pm, h \to 0.$$ (3.4)

Together with the assumption that the free energy scale as the inverse correlation volume

$$f_\infty(t, h) \sim \xi_\infty^{-d},$$ (3.5)

one has a complete description of scaling in the thermodynamic limit, from which the scaling relations (2.37)–(2.40) follow.

Firstly, Eq.(3.3) with h set to zero and Eq.(3.5) deliver the hyperscaling relation $\nu d = 2 - \alpha$. Next, differentiating Eq.(3.3) with respect to field gives $m_\infty(t) \sim |t|^{2-\alpha-\Delta}$, from which we have $\Delta = \beta + \gamma$. Combined with Eq.(3.2) this gives Widom's relation (2.39). Differentiating a second time gives $\chi_\infty(t) \sim |t|^{2-\alpha-2\Delta}$. Identifying the exponent as $-\gamma$ delivers the relation (2.38).

The starting point for the standard derivation of Fisher's relation (2.40) is the fluctuation-dissipation theorem

$$\chi_\infty(t) = \int_0^{\xi_\infty(t)} d^d\mathbf{r}\, G_\infty(r, t, 0),$$ (3.6)

having bounded the integral by the correlation length. Then, from the form (3.4),

$$\chi_\infty(t) = \int_0^{\xi_\infty(t)} dr\, r^{1-\eta} \mathcal{G}_\pm \left[\frac{r}{\xi_\infty(t)} \right].$$ (3.7)

Fixing $r/\xi_\infty(t) = x$, this gives

$$\chi_\infty(t) = \xi_\infty^{2-\eta} \int_0^1 dx\, x^{1-\eta} \mathcal{G}_\pm(x).$$ (3.8)

Finally, inserting the scaling behaviour for ξ_∞ and χ_∞, one obtains Fisher's relation[23] $\gamma = \nu(2 - \eta)$.

We will revisit this derivation in Sec.8.2 where we will see that this 50-year old scaling relation needs to be supplemented above the upper critical dimension.

The Widom scaling ansatz may be justified through Kadanoff's block spin renormalization approach.[15] One partitions the lattice into blocks of size b (in units of a) and replaces each of the b^d spins in a block by a single block spin s_I. One then rescales all lengths by an amount b. At the critical point $(t, h) = (0, 0)$ (recall $t = T/T_c - 1$ is the reduced temperature), the correlation length is infinite and remains so after the block spin transformation – it is a fixed point of the transformation which maps (t, h) to new values (t', h'). Near the critical point, the relationship between the original and renormalized parameters is $t' = \lambda_t(b)t$ and $h' = \lambda_h(b)h$. Demanding that successive blocking is equivalent to a single transformation, $\lambda_i(b_2)\lambda_i(b_1) = \lambda_i(b_1 b_2)$ and that the identity transformation effect no change, delivers the expectation that $t' = b^{y_t}t$ and $h' = b^{y_h}h$ with $y_i > 0$ for $i = t, h$. The first of these then gives $\xi'_\infty(t) \sim |t'|^{-\nu} \sim b^{-y_t\nu}\xi_\infty(t)$. But since in Kadanoff's approach the correlation length is transformed as $\xi'_\infty = \xi_\infty/b$, we identify

$$\nu = \frac{1}{y_t}. \tag{3.9}$$

Demanding that the partition function remains unchanged under the real-space renormalization transformation $Z_L(t, h) = Z_{L'}(t', h')$, where $L' = L/b$ (here, $N = L^d$ is the number of original spins and $N' = L'^d$ is the number of block spins), delivers for the free energy in the critical region,

$$f_\infty(t, h) = b^{-d}f_\infty(b^{y_t}t, b^{y_h}h). \tag{3.10}$$

This is a generalised homogeneous function. From this, Widom's scaling follows as

$$f_\infty(t, h) = |t|^{\nu d}f_\infty(\pm 1, h|t|^{-y_h/y_t}).$$

Comparing to Eq.(3.3), one has $\nu d = 2 - \alpha$ and

$$\Delta = \frac{y_h}{y_t} \quad \text{or} \quad y_h = \frac{\beta\delta}{\nu}. \tag{3.11}$$

Therefore the assumption that the renormalized partition function take the same form as the original one delivers a generalised homogeneous free energy, which then delivers Widom's ansatz and hyperscaling.

Kadanoff's block spin technique can also be applied to the correlation function. Write $m_I = b^{-d} \sum_{i \in I}^{b^d} s_i$ and define the block spin $S_I = \text{sgn}(m_I) = m_I / |m_I|$. Then,

$$
\begin{aligned}
G_\infty(r', t', h') &= \langle S_I S_J \rangle - \langle S_I \rangle \langle S_J \rangle \\
&= \frac{1}{|m_I||m_J|} \frac{1}{b^{2d}} \sum_{i \in I}^{b^d} \sum_{j \in J}^{b^d} \langle s_i s_j \rangle - \langle s_i \rangle \langle s_j \rangle \\
&= \frac{1}{|m_I||m_J|} G_\infty(r, t, h).
\end{aligned}
$$

From FSS, $m_I \sim b^{-\beta/\nu}$, so we have $G_\infty(r', t', h') \sim b^{\beta/\nu} G_\infty(r, t, h)$ or

$$
G_\infty(r, t, h) \sim b^{-\frac{\beta}{\nu}} G_\infty(rb^{-1}, tb^{y_t}, hb^{y_h}). \tag{3.12}
$$

The scaling ansatz (3.4) follows from this.

4. Mean-Field Theory, Landau Theory, ϕ^4 Theory and the Ginzburg Criterion

The mean-field theory[28,29] for the Ising model and Landau theory[30] are both historically and conceptually the basis for deeper, firmer and more realistic theories of critical phenomena. They also produce critical exponents which obey the scaling relations (except hyperscaling). Here we present both theories along with a brief account of a criterion which marks their validity.

4.1. *Mean-field theory for the Ising model*

The Ising Hamiltonian is given by Eqs.(2.2), (2.3) and (2.4),

$$
\mathcal{H} = -J \sum_{\langle i,j \rangle} s_i s_j - H \sum_i s_i. \tag{4.1}
$$

Writing $s_i = m_L + \delta s_i$ in the first term, in which $m_L = \langle s_i \rangle$,

$$
\mathcal{H} = -J \sum_{\langle i,j \rangle} (m_L + \delta s_i)(m_L + \delta s_j) - H \sum_i s_i. \tag{4.2}
$$

(In writing m_L as independent of i, we have again assumed translational invariance for simplicity.) We express this as

$$
\mathcal{H} = \mathcal{H}_{\text{MF}} + \Delta \mathcal{H}, \tag{4.3}
$$

where

$$\mathcal{H}_{\mathrm{MF}} = Jm_L^2 \sum_{\langle i,j \rangle} 1 - 2Jm_L \sum_{\langle i,j \rangle} s_i - H \sum_i s_i, \tag{4.4}$$

having reinstated s_i, and

$$\Delta\mathcal{H} = -J \sum_{\langle i,j \rangle} (\delta s_i)(\delta s_j). \tag{4.5}$$

The mean-field approximation consists of neglecting second-order fluctuations so that the energy of the model is simply $\mathcal{H}_{\mathrm{MF}}$.

Introduce the coordination number q as the number of nearest neighbours. For example, a hypercubic lattice with periodic boundary conditions has $q = 2d$. Then $\sum_{\langle i,j \rangle} 1 = qN/2$ and $\sum_{\langle i,j \rangle} s_i = (q/2) \sum_i s_i$, so that

$$\mathcal{H}_{\mathrm{MF}} = \frac{q}{2} N J m_L^2 - (Jqm_L + H) \sum_i s_i. \tag{4.6}$$

The Hamiltonian has been decoupled into a sum of single-body, non-interacting effective Hamiltonians in an effective mean-field $H_{\mathrm{eff}} = H + Jqm_L$. Now sum over the configurations and insert into the partition function to obtain

$$Z_{\mathrm{MF}} = \sum_{\{s_i\}} e^{-\beta \mathcal{H}_{\mathrm{MF}}} = e^{-\frac{\beta N q J m_L^2}{2}} [2\cosh(\beta H + J\beta q m_L)]^N. \tag{4.7}$$

From Eq.(2.5), the free energy is then

$$f_{\mathrm{MF}} = \frac{qJm_L^2}{2} - k_B T \ln[2\cosh(\beta H + J\beta q m_L)]. \tag{4.8}$$

Eq.(2.10) then gives the transcendental equation

$$m_L = \tanh[\beta(H + Jqm_L)], \tag{4.9}$$

or

$$\tanh^{-1} m_L = \frac{1}{2} \ln\left(\frac{1+m_L}{1-m_L}\right) = \beta H + \beta Jqm_L. \tag{4.10}$$

Expanding the inverse hyperbolic function,

$$m_L + \frac{m_L^3}{3} + \cdots = \beta H + \beta Jqm_L. \tag{4.11}$$

When $h = \beta H = 0$, this delivers the solutions $m_L = 0$, which can hold for any T, and $m_L = \pm\sqrt{3(\beta Jq - 1)}$, which is real only for $\beta \geq 1/Jq$. Identify

$$\beta_c = \frac{1}{qJ} \quad \text{or} \quad T_c = \frac{qJ}{k_B}. \tag{4.12}$$

There is therefore a phase transition from a zero-magnetisation phase when $H = 0$ and $T > T_c$ to a magnetised phase when $H = 0$ and $T < T_c$. These solutions coincide at T_c.

Differentiating Eq.(4.11) with respect to h gives $\chi_L(1 - \beta/\beta_c) + m_L^2 \chi_L + \cdots = \beta$, leading to $\gamma = 1$ in the thermodynamic limit. The critical isotherm $\beta = \beta_c$ gives $m_L^3 = 3\beta_c H$ so that $\delta = 3$. Finally, the internal energy is the expectation value of the Hamiltonian in Eq.(2.15). Differentiating with respect to temperature then delivers $\alpha = 0$.

The mean-field theory delivers phase transitions even for finite L and even in one dimension. However, there can be no genuine transition in a finite-size system and, according to Ising's calculation, there should be no spontaneous magnetisation in $d = 1$.[3] At the other extreme, mean-field theory becomes exact in the limit where the interactions are between all pairs of spins and not just nearest neighbours. It is also exact in the limit of infinite dimensionality. In summary, mean-field theory leads to the prediction $\alpha = 0$, $\beta = 1/2$, $\gamma = 1$ and $\delta = 3$, independent of dimensionality.

Rather than expanding out Eq.(4.9), one can expand the more fundamental equation (4.8). One finds (dropping the subscripts L),

$$f_{\mathrm{MF}} = f_0 - Hm + a_2(T - T_c)m^2 + a_4 m^4 + \ldots, \qquad (4.13)$$

where $f_0 = -k_B T \ln 2$, $a_2 = k_B/2$ and $a_4 = k_B T/12$. When $T > T_c$, the order parameter vanishes, so that $f_{\mathrm{MF}} = f_0$.

One can now proceed to determine m, χ and c in the usual manner and verify that the Taylor expansion of the mean field delivers the same scaling behaviour in the vicinity of the critical point as the full mean-field theory.

4.2. Landau theory

Eq.(4.13) coincides with Landau's phenomonological approach to phase transitions. That approach is to identify the order parameter ϕ_0 and its symmetries and then to construct a Hamiltonian from all possible invariants subject to spatial or space-time symmetries.[30] The Ising model has symmetry under $\phi_0 \to -\phi_0$, so that the polynomial should contain only even powers. The idea behind Landau's approach is that if the order parameter is small near the phase transition, the free energy can be expanded in powers of ϕ_0. That expansion can then be truncated close enough to the transition point. The coefficients of the power series are functions of the control parameters T and H. For the symmetries of the Ising model then, the first few terms are given by

$$\beta f(t,h;\phi_0) = \beta f_0(t,h) + [r_0(t,h)/2]\phi_0^2(t,h) + [u(t,h)/4]\phi_0^4(t,h) - h\phi_0(t,h).$$

Here we have expanded βf instead of f for later convenience when we connect with ϕ^4 theory. We have also introduced the expansion coefficients as fractions for the same reason. We next absorb $\beta \approx \beta_c$ into f and write this as

$$f(t,h;\phi_0) = f_0(t,h) + \frac{r_0(t,h)}{2}\phi_0^2(t,h) + \frac{u(t,h)}{4}\phi_0^4(t,h) - h\phi_0(t,h). \quad (4.14)$$

Minimising Eq.(4.14) with respect to the order parameter ϕ_0, one obtains

$$\frac{\delta f}{\delta \phi_0} = r_0(t,h)\phi_0 + u(t,h)\phi_0^3 - h = 0, \quad (4.15)$$

$$\frac{\delta^2 f}{\delta \phi_0^2} = r_0(t,h) + 3u(t,h)\phi_0^2 > 0. \quad (4.16)$$

If $h = 0$, and if $r_0(T) > 0$, Eq.(4.15) can only hold if $\phi_0 = 0$, so we identify this as the symmetric phase ($T > T_c$). If $r_0 < 0$, the equation permits the solution

$$\phi_0 = \sqrt{-\frac{r_0(T)}{u(T)}}, \quad (4.17)$$

which we can associate with the broken ($T < T_c$) phase. In both cases Eq.(4.16) is satisfied. Since $r_0(T)$ changes sign at T_c, its expansion in terms of T should take the form

$$r_0(T) = r_{01}t + r_{03}t^3 + \ldots, \quad (4.18)$$

where $r_{01} > 0$. Similarly expanding

$$u(T) = u_0 + u_1 t + \ldots, \quad (4.19)$$

with $u_0 > 0$, Eq.(4.17) becomes

$$\phi_0 = \sqrt{\frac{r_{01}}{u_0}}|t|^{\frac{1}{2}} + \ldots, \text{ for } t < 0. \quad (4.20)$$

Since ϕ_0 is the order parameter in the Landau theory, we can identify the mean-field critical exponent $\beta = 1/2$.

The specific heat is $c = -T\partial^2 f/\partial T^2 \sim -\partial^2 f/\partial t^2$. Firstly,

$$\frac{\partial f}{\partial t} = \frac{r_0'}{2}\phi_0^2 + r_0\phi_0\phi_0' + \frac{u'}{4}\phi_0^4 + u\phi_0^3\phi_0',$$

with primes indicating derivatives with respect to t. Now, the leading terms in the expansions of u and u' are constant, while ϕ_0' brings in a term proportional to $|t|^{-1}$, so the derivatives of u give sub-leading terms and can be dropped. (We can treat u as a t-independent parameter.) This gives

$$-c = \frac{\partial^2 f}{\partial t^2} = 2r_{01}\phi_0\phi_0' + r_{01}(\phi_0')^2 + r_{01}\phi_0\phi_0'' + 3u_0\phi_0^2(\phi_0')^2 + u_0\phi_0^3\phi_0''.$$

We use $\phi_0 = 0$ above T_c and Eq.(4.17) below T_c to arrive at

$$c \sim \begin{cases} 0 & \text{if } t > 0 \\ \frac{r_0^2}{2u_0} & \text{if } t < 0 \end{cases}. \tag{4.21}$$

Therefore we identify $\alpha = 0$.

If $t = 0$ or $r_0 = 0$ in Eq.(4.15), we have

$$\phi_0^3 = \frac{h}{u}, \tag{4.22}$$

yielding $\delta = 3$.

Differentiating (4.15) with respect to h and identifying the susceptibility as $\chi = \partial\phi_0/\partial h$ gives, in the absence of the external field,

$$\chi = \begin{cases} \frac{1}{r_0(t)} & \text{if } t > 0 \\ -\frac{1}{2r_0(t)} & \text{if } t < 0 \end{cases}. \tag{4.23}$$

Thus we identify $\gamma = 1$.

To obtain the correlation function, we require the Ornstein-Zernike extension[31] of the Landau theory (4.14)

$$F[\phi_0, h] = \int d^d\mathbf{r} \left[\frac{1}{2}[\nabla\phi_0(\mathbf{r})]^2 + \frac{r_0}{2}\phi_0^2(\mathbf{r}) + \frac{u}{4}\phi_0^4(\mathbf{r}) - h(\mathbf{r})\phi_0(\mathbf{r}) \right]. \tag{4.24}$$

(A general coefficient of the $(\nabla\phi_0)^2$ term here can be incorporated into the remaining coefficients. Here we set it to $1/2$ for later convenience.) Minimising,

$$\frac{\delta F}{\delta\phi_0(\mathbf{r})} = r_0\phi_0(\mathbf{r}) + u\phi_0^3(\mathbf{r}) - h(\mathbf{r}) - \nabla^2\phi_0(\mathbf{r}) = 0. \tag{4.25}$$

Differentiate the associated Eq.(4.25) with respect to field $h(\mathbf{r}')$ at location $\mathbf{r}'$ to obtain

$$r_0 G(\mathbf{r}, \mathbf{r}') + 3u\phi^2(\mathbf{r})G(\mathbf{r}, \mathbf{r}') - \nabla^2 G(\mathbf{r}, \mathbf{r}') = \delta(\mathbf{r} - \mathbf{r}'). \tag{4.26}$$

If the field h is uniform, translational invariance means that $G(\mathbf{r}, \mathbf{r}') = G(|\mathbf{r} - \mathbf{r}'|)$. We find

$$-\nabla^2 G(r) + R_0 G(r) = \delta(r), \tag{4.27}$$

where $R_0 = r_0$ if $T > T_c$, $R_0 = -2r_0$ if $T < T_c$ and $R_0 = 0$ if $T = T_c$.

The Fourier transform of Eq.(4.27) is

$$(k^2 + R_0)\tilde{G}(k) = 1. \tag{4.28}$$

The solution of Eq.(4.28) is the Ornstein-Zernike form,[31]

$$\tilde{G}(k) = \frac{1}{k^2 + \xi^{-2}}, \tag{4.29}$$

which is exact for Landau theory. Here

$$\xi = R_0^{-\frac{1}{2}} \tag{4.30}$$

is the correlation length from Eq.(2.25). The inverse Fourier transform is then

$$G(r) \sim \frac{1}{r^{d-2}} g\left(\frac{r}{\xi}\right), \tag{4.31}$$

where

$$g\left(\frac{r}{\xi}\right) = \left(\frac{r}{\xi}\right)^{\frac{d}{2}-1} K_{\frac{d}{2}-1}\left(\frac{r}{\xi}\right), \tag{4.32}$$

in which $K_{\frac{d}{2}-1}$ is a modified Bessel function. Therefore

$$G(r) \sim \frac{\xi^{1-d/2}}{r^{d/2-1}} K_{\frac{d}{2}-1}\left(\frac{r}{\xi}\right). \tag{4.33}$$

Now, $K_\nu(x) \sim 1/x^\nu$ as $x \to 0$, so when $\xi \to \infty$,

$$G(r) \sim \frac{1}{r^{d-2}} \quad \text{or} \quad \eta = 0, \tag{4.34}$$

the correlation length having dropped out. For finite $\xi \ll r$, the asymptotic behaviour $K_\nu(x) \to e^{-x}/\sqrt{x}$ for large x gives

$$G(r) \sim \frac{1}{r^{(d-1)/2}} e^{-r/\xi}. \tag{4.35}$$

4.3. *The Ginzburg-Landau-Wilson ϕ^4 Theory*

To go beyond mean-field theory or Landau theory, we need to be able to take into account fluctuations in the field $\phi(\mathbf{r})$. The connection between the Ising model and ϕ^4 theory is established through the renormalization group. The Ginzburg-Landau-Wilson partition function is

$$Z[h] = \int \mathcal{D}\phi \exp\left(-S[\phi]\right), \qquad (4.36)$$

where the action is

$$S[\phi] = \int d^d\mathbf{r} \left\{ \frac{1}{2}[\nabla\phi(\mathbf{r})]^2 + \frac{r_0}{2}\phi^2(\mathbf{r}) + \frac{u}{4}\phi^4(\mathbf{r}) - h(\mathbf{r})\phi(\mathbf{r}) \right\}, \qquad (4.37)$$

having also included a source or field $h(\mathbf{r})$. The functional integration in Eq.(4.36) is over fluctuating continuous fields ϕ. We write the Helmholtz free energy as

$$F[h] = -\ln Z[h]. \qquad (4.38)$$

Since $F[h]$ is concave, we define the convex functional

$$W[h] = -F[h] = \ln Z[h]. \qquad (4.39)$$

From Eq.(2.10), the magnetisation is

$$m(\mathbf{r}) = \frac{\delta W[h]}{\delta h(\mathbf{r})} = \langle \phi(\mathbf{r}) \rangle. \qquad (4.40)$$

Also, from Eq.(2.14), the connected correlation function is

$$G(\mathbf{r} - \mathbf{r}') = \frac{\delta^2 W}{\delta h(\mathbf{r})\delta h(\mathbf{r}')} = \frac{\delta m(\mathbf{r})}{\delta h(\mathbf{r}')}. \qquad (4.41)$$

The associated Legendre transformation

$$\Gamma[m] = \int d^d\mathbf{r} \ [m(\mathbf{r})h(\mathbf{r})] - W[h] \qquad (4.42)$$

is called the Gibbs free energy. Its usefulness is linked to the correlation function. Differentiating Eq.(4.42),

$$\frac{\delta\Gamma[m]}{\delta m(\mathbf{r})} = h(\mathbf{r}) + m\frac{\delta h}{\delta m} - \frac{\delta W}{\delta h}\frac{\delta h}{\delta m},$$

the last two terms cancel due to Eq.(4.40). Therefore

$$\frac{\delta\Gamma[m]}{\delta m(\mathbf{r})} = h(\mathbf{r}). \qquad (4.43)$$

Differentiating again,

$$\frac{\delta^2 \Gamma[m]}{\delta m(\mathbf{r}) \delta m(\mathbf{r}')} = \frac{\delta h(\mathbf{r})}{\delta m(\mathbf{r}')} = G^{-1}(\mathbf{r}, \mathbf{r}'), \qquad (4.44)$$

from Eq.(4.41).

4.4. Ginzburg criterion

The Ginzburg criterion explains the agreement between the values of the critical exponents above d_c dimensions and the mean-field predictions.[32] To discuss it, we return to the energy Eq.(4.5) which was neglected in the mean-field Hamiltonian,

$$\Delta \mathcal{H} = -J \sum_{\langle i,j \rangle} (s_i - m_L)(s_j - m_L) = -J \sum_{\langle i,j \rangle} G_L(r_i, r_j) = -\frac{Jq}{2\beta} \chi_L, \quad (4.45)$$

having used the fluctuation-dissipation theorem. The neglect of this term is justified if its contribution to the energy is small compared to that of the mean-field part. From Eq.(4.6), this is the case in zero field if $|\Delta \mathcal{H}| \ll \mathcal{H}_{\mathrm{MF}}$ or

$$\frac{\chi_L}{\beta} \ll N m_L^2. \qquad (4.46)$$

In the infinite-volume limit, the procedure is to replace N by ξ^d, so that the Ginzburg criterion becomes

$$\frac{\chi_\infty}{\beta} \ll \xi_\infty^d m_\infty^2, \qquad (4.47)$$

or $\nu d > 2\beta + \gamma = 2 - \alpha$. Using, for self-consistency, the mean-field values of the critical exponents this gives the criterion that $d > 4$. Therefore, mean-field and Landau theory should deliver meaningful results above $d_c = 4$ dimensions.

We will revisit the Ginzburg criterion in Section 10, in the light of developments outlined in the interim.

5. Wilson's Renormalization-Group Theory

Wilson's approach[33] goes beyond Kadanoff's heuristic approach in that new coupling constants can be generated with each renormalization-group transformation. It delivers an explanation for universality and the calculation of critical exponents. We require that the partition function $Z_L(\mathcal{H})$ be unchanged under the renormalization-group transformation, s.t.

$Z_{L'}(\mathcal{H}') = Z_L(\mathcal{H})$, where $L' = b^{-1}L$. In terms of the free energy, this means

$$f_{L'}(\mathcal{H}') = b^d f_L(\mathcal{H}). \tag{5.48}$$

Lengths are reduced by a factor b through the renormalization group, and the spins are also rescaled

$$s_x \to s'_{x'} = b^{d_\phi} s_x \quad \text{or} \quad \phi(z) \to \phi'(z') = b^{d_\phi}\phi(z). \tag{5.49}$$

The Hamiltonian is generally written

$$\mathcal{H} = \vec{\mu}\vec{S} = \sum_i \mu_i S_i, \tag{5.50}$$

where $\vec{\mu}$ is a vector in the space of all possible parameters which may govern the system and where $\vec{S}$ represent different interactions. In the Ising or ϕ^4 case, we may consider $\mu_1 = t$, $\mu_2 = h$, $\mu_3 = u$, etc., with $S_1 = \int d^d\mathbf{r} \; \phi^2(\mathbf{r})/2$ $S_2 = \int d^d\mathbf{r} \; \phi^4(\mathbf{r})/4$ and $S_3 = -\int d^d\mathbf{r} \; \phi(\mathbf{r})$.

The transformation maps

$$\vec{\mu} \to \vec{\mu}' = R_b\mu, \tag{5.51}$$

and fixed points are given by

$$\vec{\mu}^* = R_b\vec{\mu}^*. \tag{5.52}$$

Repeated application of the renormalization-group transformation reduces length scales by a factor of b. The correlation length remains unchanged in the thermodynamic limit only if it is infinite or zero. We expand about a fixed point (5.52),

$$\vec{\mu} = \vec{\mu}^* + \delta\vec{\mu}, \tag{5.53}$$
$$\vec{\mu}' = \vec{\mu}^* + \delta\vec{\mu}', \tag{5.54}$$

such that

$$\delta\vec{\mu}' = A_b(\vec{\mu}^*)\delta\vec{\mu}. \tag{5.55}$$

Two successive scale transformations by factors b_1 and b_2 are assumed to be equivalent to a single transformation by a scale factor of $b_1 b_2$, so that

$$A_{b_1 b_2} = A_{b_1} A_{b_2}. \tag{5.56}$$

If the eigenvalues and eigenvectors of A_b are given by

$$A_b\vec{v}_i = \lambda_i\vec{v}_i, \tag{5.57}$$

then Eq.(5.56) gives that the λ_i are homogeneous functions of b:

$$\lambda_i(b) = b^{y_i} .$$ (5.58)

Expanding $\vec{\mu}$ near $\vec{\mu}^*$ in terms of the eigenvectors $\vec{v}_i$, one has

$$\vec{\mu} = \vec{\mu}^* + \sum_i g_i \vec{v}_i .$$ (5.59)

The g_i here are called linear scaling fields. With $\delta\vec{\mu} = \sum_i g_i \vec{v}_i$, one now has

$$\delta\vec{\mu}' = \sum_i g_i' \vec{v}_i = \sum_i g_i \lambda_i \vec{v}_i$$ (5.60)

so that

$$g_i' = b^{y_i} g_i .$$ (5.61)

This gives how the linear scaling fields transform under the renormalization group and backs up the Kadanoff scaling picture.

If $\lambda_i > 0$, the scaling field g_i is augmented under renormalization group and the system is driven away from the fixed point. In this case, λ_i is called a relevant scaling variable. If $\lambda_i < 0$, the associated g_i is an irrelevant scaling field and λ_i is called an irrelevant scaling variable. If $\lambda_i = 0$, it is marginal.

We re-express Eq.(5.48) in terms of the linear scaling fields as

$$f_\infty(\vec{\mu}) = b^{-d} f_\infty(\vec{\mu}') .$$ (5.62)

Near a fixed point, then, where $\vec{\mu}$ and $\vec{\mu}'$ may be written in terms of linear scaling fields, this may again be rewritten as

$$f_\infty(g_1, g_2, g_3, \dots) = b^{-d} f_\infty (b^{y_1} g_1, b^{y_2} g_2, b^{y_3} g_3, \dots) .$$ (5.63)

For the ϕ^4 theory, the renormalization-group approach leads to two fixed points. The Wilson-Fisher fixed point is stable below d_c and the Gaussian fixed point is stable above d_c. Eq.(5.61) reads

$$r_0 \to r_0' = b^{y_t} r_0,$$ (5.64)

$$h \to h' = b^{y_h} h,$$ (5.65)

$$u \to u' = b^{y_u} u .$$ (5.66)

The first of these may also be written $\to t' = b^{y_t} t$, after Eq.(4.18). Eq.(5.63) for the ϕ^4 theory is then

$$f_\infty(t, h, u) = b^{-d} f_\infty (b^{y_t} t, b^{y_h} h, b^{y_u} u) .$$ (5.67)

This is the *scaling hypothesis* (3.10) with the irrelevant field accounted for.

Differentiating Eq.(5.67) appropriately to obtain the thermodynamic functions, one finds

$$\alpha = 2 - \frac{d}{y_t}, \quad \beta = \frac{d - y_h}{y_t}, \quad \frac{1}{\delta} = \frac{d}{y_h} - 1, \quad \gamma = \frac{2y_h - d}{y_t}. \tag{5.68}$$

A similar form for the correlation function,

$$G_\infty(r, t, h, u) = b^{-2d_\phi} G_\infty(b^{-1}r, b^{y_t}t, b^{y_h}h, b^{y_u}u), \tag{5.69}$$

in which d_ϕ is the scaling dimension of the fields defined in Eq(5.49), delivers

$$G_\infty(r, t, 0, 0) = b^{-2d_\phi} G_\infty(b^{-1}r, b^{y_t}t, 0, 0), \tag{5.70}$$

having set both h and u to zero. Then setting $b = r$, one finds

$$G_\infty(r, t, 0, 0) = \frac{g(r/t^{-\frac{1}{y_t}})}{r^{2d_\phi}}, \tag{5.71}$$

for some function g. Comparing to the general form $G_\infty(r, t) = \exp -(r/\xi)/r^{d-2+\eta}$, one concludes

$$\nu = \frac{1}{y_t}, \tag{5.72}$$

$$\eta = 2d_\phi + 2 - d. \tag{5.73}$$

These recover identities established in Kadanoff's approach (see Sec. 3). Alternatively, the scaling form

$$\xi_\infty(t, h, u) = b\xi_\infty(b^{y_t}t, b^{y_h}h, b^{y_u}u) \tag{5.74}$$

directly delivers Eq.(5.72).

5.1. *Scaling at the Gaussian fixed point*

At the Gaussian fixed point, the renormalization-group scaling dimensions are obtained by power counting: For the action (4.37) to remain dimensionless under the transformation $r \to b^{-1}r$, $\phi \to b^{d_\phi}\phi$, $r_0 \to b^{y_t}r_0$, $u \to b^{y_u}u$, $h \to b^{y_h}h$, one requires

$$d_\phi = \frac{d}{2} - 1, \tag{5.75}$$

$$y_t = 2, \tag{5.76}$$

$$y_u = 4 - d, \tag{5.77}$$

$$y_h = \frac{d}{2} + 1. \tag{5.78}$$

Inserting these into Eqs.(5.68) and (5.72), we identify

$$\alpha = 2 - \frac{d}{2}, \quad \beta = \frac{d}{4} - \frac{1}{2}, \quad \gamma = 1, \quad \frac{1}{\delta} = \frac{d-2}{d+2}, \quad \nu = \frac{1}{2}, \quad (5.79)$$

with $\eta = 0$ from Eq.(5.73).

According to renormalization-group theory, the Gaussian fixed point is stable at and above d_c, so these values are supposed to be valid there. Therefore they should coincide with the results of Landau theory, $\alpha = 0$, $\beta = 1/2$, $\gamma = 1$, $\delta = 1/3$, $\nu = 1/2$ and $\eta = 0$. We see that, while γ, ν and η are in agreement, the values for α, β and δ disagree, except at $d = d_c$ itself.

6. Dangerous Irrelevant Variables

To repair the shortcomings identified above the upper critical dimension, Fisher introduced the notion of dangerous irrelevant variables.[34] The danger should apply to the free energy because it is associated with α, β and δ. However, since the values of ν and η coming from the Gaussian fixed point are correct (they coincide with Landau theory), one does not expect danger for either the correlation function or the correlation length. Moreover, since the susceptibility is linked to the correlation function by the fluctuation-dissipation theorem, one expects no danger for χ either. Indeed, the value $\gamma = 1$ coming from the Gaussian fixed point is the same as that from mean-field theory.

6.1. *The Thermodynamic Limit*

Eq.(4.21) shows that the mean-field specific heat behaves as u^{-1} (in the broken symmetry phase). Therefore the naive process of setting u to zero, or ignoring its role in the free energy derivatives, is incorrect. In fact, the second t-derivative $f_{tt}(x, y, z)$ should scale as z^{-1} for small values of the third argument. This identifies the variable u as dangerous in the ϕ^4 theory; it cannot be set to zero. From Eqs.(4.21) and (5.67), then, one expects

$$c_\infty(t, 0, u) \sim b^{-d+2y_t-y_u} u^{-1} \tilde{f}_{tt}(b^{y_t} t). \quad (6.1)$$

Now setting $b = |t|^{-1/y_t}$, one has

$$c_\infty(t, 0, u) \sim |t|^{-\frac{d-2y_t+y_u}{y_t}} \tilde{f}_{tt}(\pm 1). \quad (6.2)$$

Similarly, Eq.(4.17) gives that the mean-field spontaneous magnetisation behaves with u as $u^{-1/2}$. Therefore, in the mean-field case the first h-derivative $f_h(x, y, z)$ behaves as $z^{-1/2}$ for small z, so that z cannot simply be set to 0. Instead,

$$m_\infty(t, 0, u) \sim b^{-d+y_h-\frac{1}{2}y_u} u^{-\frac{1}{2}} \tilde{f}_h\left(b^{y_t} t\right). \tag{6.3}$$

Again setting $b = t^{-1/y_t}$, one obtains

$$m_\infty(t, 0, u) \sim t^{\frac{d-y_h+\frac{1}{2}y_u}{y_t}}. \tag{6.4}$$

For the critical isotherm, the role of the dangerous irrelevant variable is apparent from Eq.(4.22), which indicates that $f_h(0, y, z)$ behaves as $z^{-1/3}$. Then

$$m_\infty(0, h, u) \sim b^{-d+y_h-\frac{1}{3}y_u} \tilde{f}_h\left(0, b^{y_h} h, b^{y_u} u\right), \tag{6.5}$$

which leads to

$$m_\infty(0, h, u) \sim h^{\frac{d-y_h+\frac{1}{3}y_u}{y_h}}. \tag{6.6}$$

Finally, since from Eq.(4.23) the leading mean-field susceptibility does not depend on u, we may infer that u is not dangerous for χ, as noticed above. Therefore Eq.(4.23) is expected to hold. Similar statements hold for the correlation function and correlation length.

From these considerations, one identifies

$$\alpha = -\frac{d - 2y_t + y_u}{y_t}, \tag{6.7}$$

$$\beta = \frac{d - y_h + \frac{1}{2}y_u}{y_t}, \tag{6.8}$$

$$\gamma = \frac{2y_h - d}{y_t}, \tag{6.9}$$

$$\frac{1}{\delta} = \frac{d - y_h + \frac{1}{3}y_u}{y_h}. \tag{6.10}$$

Eqs.(5.72) and (5.73) are expected to remain valid for ν and η since neither the correlation length nor the correlation function are expected to be affected by the danger of u. Finally, with the scaling dimensions (5.75)–(5.76) from power counting, one obtains the correct exponents

$$\alpha = 0, \quad \beta = \frac{1}{2}, \quad \gamma = 1, \quad \delta = 3, \tag{6.11}$$

along with

$$\nu = \frac{1}{2}, \quad \eta = 0. \tag{6.12}$$

6.2. Finite-size scaling: Naive approach above d_c

Having using dangerous irrelevant variables to repair scaling in the thermodynamic limit above the upper critical dimension, we now turn to FSS there. A naive application of the FSS ansatz (2.27) replaces $\xi_\infty(t)$ by L and one finds

$$c_L \sim L^0, \quad m_L \sim L^{-1}, \quad \chi_L \sim L^2, \quad \xi_L \sim L. \tag{6.13}$$

We refer to this *Gaussian FSS* or *Landau FSS* because it comes from applying the traditional FSS ansatz (2.27) to the mean-field exponents.

Eq.(6.13) is, however in disagreement with an explicit analytical calculation by Brézin for the n-vector model with periodic boundary conditions (PBC's).[9] There have also been many numerical studies throughout the years[35–52] which confirm that Eqs.(6.13) for m_L and for χ_L do not hold for the Ising model when PBC's are used. To understand how the conventional paradigm deals with this inconsistency, we turn to the Gaussian model.

6.3. FSS in the Gaussian model with periodic boundaries

The Gaussian or free field theory has an action given by Eq.(4.37) dropping the quartic self-interaction term. Defined on a finite-size lattice in momentum space in vanishing field, it is given as

$$S_L^{(0)}(\phi) = \frac{1}{2} \sum_k (k^2 + r_0^2)|\hat{\phi}_k|^2, \tag{6.14}$$

where $\hat{\phi}_k$ are the Fourier-transformed fields.

The correlation function is given through Eq.(4.44) by the inverse of the quadratic part of the action,

$$\tilde{G}^{(0)}(k) = \langle \hat{\phi}_k \hat{\phi}_{-k} \rangle = \frac{1}{k^2 + r_0}. \tag{6.15}$$

Here we have omitted the subscript L from G, but it is understood that we are still considering a finite lattice. Therefore, from the fluctuation dissipation theorem,

$$\chi_L = \tilde{G}^{(0)}(0) = r_0^{-1}. \tag{6.16}$$

For PBC's, the wave vectors are

$$k_\mu = \frac{2\pi n}{La}, \tag{6.17}$$

in which $n = 1, 2, \ldots L$ and $\mu = 1, \ldots, d$. The correlation length can be defined as a second moment:

$$\xi_L = \frac{1}{k_{\min}} \left[\frac{1}{2d} \frac{\tilde{G}(0) - \tilde{G}(k_{\min})}{\tilde{G}(k_{\min})} \right]^{\frac{1}{2}} = r_0^{-\frac{1}{2}}, \tag{6.18}$$

where $k_{\min} = 2\pi/La$. Therefore both the susceptibility and the correlation length diverge at $r_0 = 0$ in the Gaussian model, *even for a finite lattice.*

This finite-size divergence is problematic. To deal with it we examine the full ϕ^4 action (4.37), which in Fourier space is

$$S_L^{\mathrm{GLW}}[\phi] = \frac{1}{2} \sum_k \left[k^2 + r_0 \right] \hat{\phi}_k \hat{\phi}_{-k} + \frac{u}{4} \frac{1}{L^d} \sum_{k_1, k_2, k_3} \hat{\phi}_{k_1} \hat{\phi}_{k_2} \hat{\phi}_{k_3} \hat{\phi}_{-k_1-k_2-k_3}. \tag{6.19}$$

We gather the quadratic terms in the zero modes,

$$S_L^{\mathrm{GLW}}[\phi] = \frac{1}{2} \sum_k \left[k^2 + r_0 + \frac{3u}{2L^d} \hat{\phi}_0^2 \right] \hat{\phi}_k \hat{\phi}_{-k} + \frac{u}{4L^d} \hat{\phi}_0^4 +$$
$$\frac{u}{4} \frac{1}{L^d} \sideset{}{'}\sum_{k_1, k_2, k_3} \hat{\phi}_{k_1} \hat{\phi}_{k_2} \hat{\phi}_{k_3} \hat{\phi}_{-k_1-k_2-k_3}, \tag{6.20}$$

where the prime indicates that the summation omits terms in which pairs of momenta vanish. Again identifying the correlation function as the inverse of the quadratic part,

$$\tilde{G}^{(0)}(k) = \langle \hat{\phi}_k \hat{\phi}_{-k} \rangle = \frac{1}{r_0 + k^2 + \frac{3u}{2L^d} \langle \hat{\phi}_0^2 \rangle}. \tag{6.21}$$

Therefore

$$\tilde{G}^{(0)}(0) = \frac{1}{r_0 + \frac{3u}{2L^d} \tilde{G}^{(0)}(0)}. \tag{6.22}$$

Solving for $\tilde{G}^{(0)}(0)$ when $r_0 = 0$, we obtain

$$\chi_L = G^{(0)}(0) = \langle \hat{\phi}_0^2 \rangle = \sqrt{\frac{2}{3u}} L^{\frac{d}{2}}, \tag{6.23}$$

which is now finite. This formula has been verified many times for finite-sized Ising systems with PBC's.[9,35-52,52]

The propagator is then

$$G^{(0)}(k) = \langle \phi_k \phi_{-k} \rangle = \frac{1}{r_0 + k^2 + \sqrt{\frac{3u}{2}} L^{-d/2}}. \tag{6.24}$$

The correlation length is

$$\xi_L^{(2)} = \frac{1}{k_{\min}} \left[\frac{1}{2d} \frac{\tilde{G}^{(0)}(0) - \tilde{G}^{(0)}(k_{\min})}{\tilde{G}^{(0)}(k_{\min})} \right]^{\frac{1}{2}} =$$

$$\frac{1}{k_{\min}} \left[\frac{1}{2d} \sqrt{\frac{2}{3u}} L^{\frac{d}{2}} k_{\min}^2 + \frac{1}{2d} - 1 \right]^{\frac{1}{2}}, \qquad (6.25)$$

when $r_0 = 0$. Now, $k_{\min} = 2\pi/La$ from Eq.(6.17). Therefore $\xi_L(0) \sim L(L^{d/2-2} + \text{const.})^{1/2}$, so that

$$\xi_L(0) \sim \begin{cases} L & \text{if } d > 4 \\ \\ L^{d/4} & \text{if } d > 4 \end{cases} . \qquad (6.26)$$

Therefore both ξ_L and χ_L remain finite at $r_0 = 0$ ($T = T_c$).

The result $\xi_L \sim L^{d/4}$ was also obtained on a theoretical basis for the large-n vector model in Ref. [9]. However, it violates an expectation that the correlation length be bounded by the length.[36] Nonetheless, it has been verified numerically for Ising systems with periodic boundaries in Refs. [47, 53,54]. See also Refs. [55–57] for studies of spin-glass systems.

We next move on to the general FSS scheme above the upper critical dimension with dangerous irrelevant variables. A question to keep in mind is whether the FSS behaviour outlined here is captured by the general scheme. We will see it that, in its original formulation, it is not. This forced the introduction of another length scale (dubbed *thermodynamic length*) to control FSS above d_c.

6.4. *FSS with dangerous irrelevant variables*

In 1985, Binder, Nauenberg, Privman and Young extended Fisher's concept of dangerous irrelevant variables to the finite-volume case.[36] We follow and assume that the finite-size counterpart of Eq.(5.67) is

$$f_L(t, h, u) = b^{-d} f_{L/b} (tb^{y_t}, hb^{y_h}, ub^{y_u}) . \qquad (6.27)$$

We make the further assumption that, for small x_3,

$$f_{L/b}(x_1, x_2, x_3) = x_3^{p_1} \bar{f}_{L/b} (x_1 x_3^{p_2}, x_2 x_3^{p_3}) . \qquad (6.28)$$

Under this assumption, the free energy may be written

$$f_L(t, h, u) = b^{-d^*} \bar{f}_{L/b} \left(tL^{y_t^*}, hL^{y_h^*} \right) , \qquad (6.29)$$

in which

$$d^* = d - p_1 y_u,$$ (6.30)

$$y_t^* = y_t + p_2 y_u,$$ (6.31)

$$y_h^* = y_h + p_3 y_u.$$ (6.32)

Here y_t^* and y_h^* effective exponents. The infinite volume limit of Eq.(6.29) yields

$$f_\infty(t, h) = t^{\frac{d^*}{y_t^*}} Y_\pm \left(h t^{-\Delta} \right),$$ (6.33)

in which

$$\Delta = \frac{y_h^*}{y_t^*}.$$ (6.34)

Now we take derivatives to obtain the thermodynamic functions. These give

$$\alpha = 2 - \frac{d^*}{y_t^*},$$ (6.35)

$$\beta = \frac{d^* - y_h^*}{y_t^*},$$ (6.36)

$$\gamma = \frac{2y_h^* - d^*}{y_t^*},$$ (6.37)

$$\frac{1}{\delta} = \frac{d^*}{y_h^*} - 1,$$ (6.38)

which in turn yield

$$y_t^* = \frac{d^*}{2 - \alpha} = \frac{d^*}{\gamma + 2\beta},$$ (6.39)

$$y_h^* = \frac{d^* \delta}{1 + \delta} = \frac{d^*(\gamma + \beta)}{\gamma + 2\beta},$$ (6.40)

$$\Delta = \beta + \gamma.$$ (6.41)

The last of these establishes Δ as the gap exponent. Also, it is at this point that we obtain the static scaling relations

$$\alpha + 2\beta + \gamma = 2, \quad \text{and} \quad \beta(\delta - 1) = \gamma.$$

Inserting the mean-field values $\alpha = 0$, $\beta = 1/2$, $\gamma = 1$ and $\delta = 1/3$, we obtain

$$y_t^* = \frac{d^*}{2}, \quad \text{and} \quad y_h^* = \frac{3d^*}{4}.$$ (6.42)

In Ref. [36] three arguments are given for the coincidence of d^* and d. This equality corresponds to $p_1 = 0$. (We revisit this in Sec. 6.5.) In this case, one has

$$y_t^* = \frac{d}{2} \quad \text{and} \quad y_h^* = \frac{3d}{4}. \tag{6.43}$$

Moreover,

$$p_2 = -\frac{1}{2} \quad \text{and} \quad p_3 = -\frac{1}{4}. \tag{6.44}$$

We can now determine the FSS of the thermodynamic functions by appropriate differentiation of Eq.(6.29). Differentiating with respect to h, setting $t = h = 0$ and $b = L$, one obtains $m_L(0,0) \sim L^{-d+y_h^*}$ and $\chi_L(0,0) \sim L^{-d+2y_h^*}$. From Eq.(6.43) then,

$$m_L \sim L^{-\frac{d}{4}}, \quad \text{and} \quad \chi_L \sim L^{\frac{d}{2}}. \tag{6.45}$$

Similarly differentiating Eq.(6.29) with respect to t one obtains $c_L(0,0) \sim L^{-d+2y_t^*}$ or

$$c_L \sim L^0. \tag{6.46}$$

Eqs.(6.45) are different from the naive Landau FSS results of Eqs.(6.13). Eqs.(6.45) are, however, in agreement with the result (6.23) for the model with PBC's. Indeed, many numerical studies using PBC's throughout the years have verified that Eqs. (6.45) give the correct FSS above the upper critical dimension. However, widespread belief is that Landau FSS holds for free boundary conditions (FBC's), at least at the critical point.

In Ref. [36], an extension of the above scenario to the correlation length or correlation function was not fully considered. The equivalent form to Eq.(6.27) for the correlation length is

$$\xi_L(t, h, u) = b\xi_{L/b}\left(tb^{y_t}, hb^{y_h}, ub^{y_u}\right). \tag{6.47}$$

Following a similar procedure to before, we write

$$\xi_L(t, h, u) = b^{1+q_1 y_u}\bar{\xi}_{L/b}\left(tb^{y_t+q_2 y_u}, hb^{y_h+q_3 y_u}\right), \tag{6.48}$$

to obtain

$$\xi_L(t, h) = L^q\bar{\xi}_1\left(tL^{y_t^{**}}, hL^{y_h^{**}}\right), \tag{6.49}$$

in which

$$q = 1 + q_1 y_u, \tag{6.50}$$

$$y_t^{**} = y_t + q_2 y_u, \tag{6.51}$$

$$y_h^{**} = y_h + q_3 y_u. \tag{6.52}$$

Setting $t = h = 0$ gives a correlation length which scales algebraically with L. However, believing that the "finite-size correlation length ξ_L is bounded by the length L", in Ref. [36] it was assumed $q_1 = 0$ so that $\xi_L \sim L$ for periodic as well as free boundaries.

The FSS prescription (2.27) therefore fails above $d = d_c = 4$. This led Binder to consider alternatives and in 1985 he introduced the *thermodynamic length*.[35] Our next step is to summarise these arguments.

6.5. *The thermodynamic length*

In order to repair FSS, Binder introduced the concept of *thermodynamic length*. In the $T < T_c$ phase, the probability density of the magnetisation M for a finite system may be approximated by the sum of two Gaussians.[58] The Gaussians are centred around the (infinite-volume) spontaneous magnetisation $\langle M \rangle = N m_{\text{sp}} = N m_\infty(0,0)$ and are of width $\sqrt{\langle (M - \langle M \rangle)^2 \rangle} = N \chi_\infty(t,0)/\beta$. The arguments of the Gaussians are therefore

$$\frac{N(m \pm m_{\text{sp}})^2}{2 k_B T \chi_\infty} = \frac{m_{\text{sp}}^2 (1 \pm m/m_{\text{sp}})^2}{2 k_B T \chi_\infty} L^d = \frac{1}{2 k_B T} \left(1 \pm \frac{m}{m_{\text{sp}}} \right)^2 \left(\frac{L}{\ell_\infty} \right)^d,$$

in which

$$\ell_\infty = \left(\frac{\chi_\infty}{m_{\text{sp}}^2} \right)^{\frac{1}{d}}. \tag{6.53}$$

Now, with $\chi_\infty(t) \sim |t|^{-1}$ and $m_\infty(t) \sim |t|^{1/2}$, we have

$$\ell_\infty(t) \sim |t|^{-\frac{2}{d}}. \tag{6.54}$$

This is called the *thermodynamic length* because it appears in the thermodynamic functions.

The assumption is that $\ell_\infty(t)$ governs FSS instead of $\xi_\infty(t)$ (which scales as $t^{-1/2}$). Indeed, the FSS ansatz (2.27) is then replaced by

$$\frac{P_L(t)}{P_\infty(t)} = \mathcal{F}_P \left[\frac{L}{\ell_\infty(t)} \right]. \tag{6.55}$$

Applied to the magnetisation, for example, this ansatz gives

$$m_L(0) = m_\infty(t) \mathcal{F}_m \left(\frac{L}{t^{-2/d}} \right) \sim L^{-\frac{d}{4}},$$

in agreement with Eq.(6.45). We can check that the new ansatz also delivers FSS for the susceptibility in Eq.(6.45).

This set-up is in accordance with the change in the homogeneity assumption from tb^{y_t} in Eq.(6.27) to the combination $tb^{y_t^*} = [b/t^{-1/y_t^*}]^{y_t^*} = [b/\ell_\infty(t)]^{y_t^*}$ provided that $\ell_\infty(t) \sim |t|^{-1/y_t^*}$. Comparing with Eq.(6.54), one has $y_t^* = d/2$. This justifies the identification of d^* with d leading to Eq.(6.43).

One notes that this profoundly modifies Fisher's original (infinite-volume) dangerous-irrelevant-variables mechanism of Sec. 6.1 because, not only the prefactor of the scaling function is altered in the $u \to 0$ limit, but also its arguments.

The finite-size counterpart of the thermodynamic length ℓ_∞ was termed *coherence length* ℓ_L in Ref. [59] and it scales as the system extent L. A so-called *characteristic length* $\lambda_L(t)$ was also introduced, as the FSS counterpart of the infinite-volume correlation length ξ_∞.

The picture set out above was, until recently, essentially the basis for the standard understanding of scaling and FSS above the upper critical dimension. The thermodynamic length $\ell_\infty(t)$ is supposed to replace $\xi_\infty(t)$ in the FSS scaling ansatz (2.27), so that

$$\frac{P_L(t)}{P_\infty(t)} = \mathcal{F}_P\left[\frac{L}{\ell_\infty(t)}\right]. \tag{6.56}$$

Although all of this delivers the correct values for the exponents above d_c, it is not satisfactory. Along with the lattice spacing a this means a number of length scales have entered the game. There have been many instances of proliferation in science which signalled a flaw in a standing paradigm and we are reminded of the quote in the Introduction. We next outline more concrete reasons for dissatisfaction before introducing the new picture.

7. An Unsatisfactory Paradigm

To summarise, dangerous irrelevant variables play a crucial role in reconciling scaling above the upper critical dimension with mean-field and Landau theory in the thermodynamic limit, where they alter the prefactor of some of the scaling functions. A naive approach to FSS then delivers Gaussian or Landau FSS in which $\chi_L \sim L^{\gamma/\nu} = L^2$ and $\xi_L \sim L$. These are incompatible, however, with exact calculations in the Gaussian model, the n-vector model and with the results of numerical simulations with PBC's. To repair this, a modification of the role of dangerous irrelevant variables was introduced whereby they also alter the arguments of the scaling functions for finite-size systems. This led to the introduction of a new length scale – the thermodynamic length – which takes over from the correlation length in the finite-size scaling ansatz.

The resulting predictions for m_L, χ_L and c_L given by Eqs.(6.45) and (6.46) have been confirmed many times over using numerical simulations of the Ising model with PBC's in five,[35–40,42–47] six,[48–50] seven[51,52] and eight[52] dimensions.

In contrast to the PBC case, however, there have been few studies of systems with FBC's above d_c.[53,54,60,61] The standard picture is that the FBC case is governed by Gaussian FSS at the critical point, in which $\chi_L \sim L^2$ due to a belief that the second expression in Eq.(6.45) "cannot hold for FBC's because it lies above a strict upper bound"[60] (namely $L^{\gamma/\nu} = L^2$) established in Ref. [62] (see also Ref. [63]). A recent numerical study of the five-dimensional Ising model with FBC's supports Gaussian FSS $\chi_L \sim L^2$ at the critical point T_c.[61]

However, a number of unsettling issues with the standard picture arise.

Firstly, the mechanism outlined in Sec. 6.4 and the FSS ansatz (6.56) do not explicitly distinguish between different sets of boundary conditions. In this sense the origin of the disparity between FSS with PBC's and FBC's is unclear. Indeed, although the Fourier analysis of Ref. [60] delivers $\chi_L \sim L^2$ for FBC's and $\chi_L \sim L^{d/2}$ for PBC's at T_c for the Ising case, it also delivers the latter result for FBC's – but only in a region away from T_c.

It also remains unexplained why the dangerous irrelevant variables mechanism affects the free energy but not the correlation function or the correlation length in the PBC case. This is especially puzzling if the arguments of the thermodynamic functions are affected as well as the prefactors, because the same arguments enter into all three functions.

Brézin[9] established that in the large-n limit of the n-vector model the correlation length for the finite system scales as $\xi_L \sim L^{d/4}$ for $d > 4$. In the $d = 4$ case, the corresponding FSS is $\xi_L \sim L(\ln L)^{1/4}$. He argued for the same behaviour in the finite-n case. In Ref. [64], an alternative ansatz to (2.27) was introduced to deal with the case of logarithmic corrections at the critical dimension itself:

$$\frac{P_L(t)}{P_\infty(t)} = \mathcal{F}_P \left[\frac{\xi_L(t)}{\xi_\infty(t)} \right]. \tag{7.1}$$

In the $d = 4$ case, the scaling dimension y_u vanishes and u is marginal rather than irrelevant. The ansatz (7.1) is therefore not dependent on the danger of u. For the Ising model or ϕ^4 theory in $d = 4$ dimensions, it gives the correct FSS for the thermodynamic functions and partition function zeros,[64–66] provided Eq.(2.52) holds with Brézin's prediction $\hat{\varphi} = 1/4$. It also delivers the correct FSS above d_c in the PBC case with $\xi_L \sim L^{d/4}$. The ansatz (7.1) reduces to Eq.(2.27) in the case where $\xi_L \sim L$. However, it remains distinct from Eq.(6.56).

Thus the principle is established that the correlation length can, in fact, exceed the length of the system, in contrast to the assumptions of Sec. 6.4 and of Ref. [36]. This was reaffirmed in Ref. [47] for PBC's in a numerical simulation of the $d = 4$ and $d = 5$ Ising models. In particular, this confirmed the FSS of the correlation length as $\xi_L \sim L^{d/4}$.

The less frequently studied case of FBC's was described in Ref. [47] as "poorly understood". Indeed, it is stated there that for FBC's "it seems obvious that even for $d > 4$ the behavior of the system will be affected when ξ_L becomes of order L", rather than the larger $L^{d/4}$. On this basis, it was believed that the standard (Gaussian) FSS expressions (corresponding to $\xi_L \sim L$) are expected to apply. The larger length scale $L^{d/4}$ was then expected to contribute to corrections to scaling in some manner which was unspecified.

7.1. Fisher's scaling relation for finite-size systems

There is another problem with the conventional paradigm. To explain it, we return to the derivation of Fisher's scaling relation in Section 3. The expression equivalent to Eq.(3.6) for a finite lattice system is

$$\chi_L(t) = \int_a^L dr\, r^{d-1} G_L(r, t), \tag{7.2}$$

in which a is the lattice spacing and we have dropped explicit dependency on h which is set to vanish. Then, from the form (3.4),

$$\chi_L = \int_0^L dr\, r^{1-\eta} \mathcal{G}_\pm\left(\frac{r}{\xi_L}\right), \tag{7.3}$$

assuming that the lower integral limit in Eq.(7.2) only delivers corrections to scaling. Fixing $r/\xi_L(t) = x$, this gives

$$\chi_L = \xi_L^{2-\eta} \int_0^{L/\xi_L} dx\, x^{1-\eta} \mathcal{G}_\pm(x). \tag{7.4}$$

When $d < d_c$, where $\xi_L \sim L$, this delivers Fisher's scaling relation

$$\frac{\gamma}{\nu} = 2 - \eta.$$

Above d_c, however, Eq.(7.4) runs into trouble. According to the conventional paradigm, $\chi_L \sim L^{d/2}$, at least for systems with PBC's. With $\xi_L \sim L$, Eq.(7.4) leads to

$$\eta = 2 - \frac{d}{2}.$$

This corresponds to a negative anomalous dimension above d_c, in disagreement with Landau theory, for which $\eta = 0$.

The negativity of the anomalous dimension was already noticed in a numerical study by Nagle and Bonner in Ref. [67]. Baker and Golner also determined a negative η in an analytical study of an Ising model for which scaling is exact.[68] Their explanation was that there are two long-range distance scales: long long range and short long range. They posited that long long-range order is controlled by a new, *different* anomalous dimension, while "short long-range order" is controlled by the usual η. The long long-range exponent *fails to satisfy the scaling relation for the anomalous dimension above the upper critical dimension.*

Luijten and Blöte revisited the problem using the relationship (2.14) to obtain the correlation function from the free energy by differentiation with respect to two local fields,[41,43]

$$\langle \phi(0)\phi(r) \rangle \sim L^{-d} \frac{\partial^2 f(t, h, u)}{\partial h(0) \partial h(r)}. \tag{7.5}$$

Using Eq.(6.27) and ignoring the danger of u, one finds $G_L \propto L^{2(y_h - d)} = L^{-(d-2)}$, which is the standard, Landau result with $\eta = 0$. However, taking account of the dangerous irrelevancy by differentiating Eq.(6.29) instead, Luijten and Blöte obtained $G_L \propto L^{2(y_h^* - d)} = L^{-d/2}$, corresponding to an anomalous dimension $\eta^* \equiv 2 - d/2$.

The problem with this interpretation is that under the standard paradigm, although u is dangerous for the free energy, it is not supposed to be dangerous for the correlation function. The approach outlined here skirts the issue by appealing to the free energy rather than directly to the correlation function.

Luijten and Blöte give a second interpretation to their anomalous dimensions.[41,43] Eq.(6.23) gives the inverse correlation function at criticality in momentum space as

$$G^{-1}(k) \sim k^2 + \sqrt{\frac{3u}{2}} L^{-\frac{d}{2}}.$$

If we identify $L^{-1} = k_{\min}$ as a momentum, we have two different decay modes. We compare these to the general form $G^{-1}(q) = q^{2-\eta}$. If we again ignore the danger and set $u = 0$, we obtain $\eta = 0$. When k is large compared to $L^{-1} = k_{\min}$ (i.e., over distances significantly shorter than the lattice extent), we obtain the same result. If, on the other hand, we acknowledge the danger and keep the u term, it delivers an anomalous dimension $\eta^* = 2 - d/2$.

So the interpretation Refs. [41,43] is that the short long-distance behaviour is governed by the exponent $\eta = 0$, which remains unaffected by the morphing of y_h into y_h^*. Long long-distance behaviour is then supposed to be ruled by η^*.

These interpretations, however, do not explain how $\eta^* = 2 - d/2$ for long long distance is manifest as $\eta = 0$ in the infinite-volume limit where field-theoretic theorems outlawing negative anomalous dimensions apply.[23,25,69] Nor do they explain why it is the supposedly long long-range anomalous dimension η^* associated with the correct dangerous-irrelevant-variables mechanism which conflicts with Landau and mean-field theory, fails to satisfy Fisher's relation and violates field theory. One would rather expect the disparity to be associated with the incorrect neglect of dangerous irrelevant variables or taking short, rather than long, long distances. Therefore *the standard paradigm does not explain scaling above the upper critical dimension.*

To summarise, the standard paradigm for scaling and FSS above the upper critical dimension is assembled in a rather ad hoc fashion. It posits that FSS becomes non-universal above d_c, where hyperscaling breaks down and extra length scales appear. Nonetheless, the numerical results were interpreted as in agreement with the detail of the theory and critical exponents are known. Thus we return to the statement by Binder et al. in quoted in Sec.1:[1] "although ... all exponents are known, ... the existing theories clearly are not so good".

8. A New Paradigm: Hyperscaling above the Upper Critical Dimension and a Negative Anomalous Dimension

We next present a recently developed picture,[53,54,70] starting with a derivation which illustrates the need for a new theory.

8.1. Self-Consistency: the Requirement for a New Universal Scaling Exponent

Here we explore the self-consistency of the FSS ansätze (6.56) and (7.1). Each reduces to the form (2.27) when $d < d_c$.

For finite Ising systems the Lee-Yang zeros form a discrete set on the imaginary-h axis.[5] We label them $h_j(L, t)$, with j indicating relative positions according to distance from the real-h axis. Since the total number of zeros is proportional to the volume L^d, these are expected to scale as a function of the ratio j/L^d, at least for sufficiently large j.[71,72] (In the case of $d = d_c = 4$ dimensions, logarithmic corrections do not enter this ratio.[27]) The counterpart of h_1 in the infinite-L limit is the Lee-Yang edge $h_{YL}(t)$. Following Ref. [27], the finite-size susceptibility is related to the Lee-Yang zeros as

$$\chi_L \sim L^{-d} \sum_{j=1}^{L^d} h_j^{-2}(L). \tag{8.1}$$

This holds *irrespective of boundary conditions*.

We first consider the standard paradigm and the ansatz (6.56) corresponding to the thermodynamic-length formalism, and $\ell_\infty(t) \sim |t|^{-2/d}$ from Eq.(6.54). The thermodynamic length, being defined for infinite volume, is not affected by boundary conditions and the ansatz gives $h_j(L) \sim (j/L^d)^{\Delta/2}$. Eq.(8.1) is then

$$\chi_L \sim L^{d(\Delta-1)} \sum_{j=1}^{L^d} j^{-\Delta}.$$

In the PBC case, the left hand side of this equation is $\chi_L \sim L^{d/2}$. Matching with the right hand side, one obtains $\Delta = 3/2$, which is, indeed, correct. However, if $\chi_L \sim L^2$ (the FBC case), one arrives at $\Delta = 1 + 2/d$. This is incorrect, except when $d = 4$. Therefore the standard paradigm leads to an *inconsistency* when the boundary conditions are free.

We next consider the ansatz (7.1) and we allow

$$\xi_L \sim L^{\mathsf{Q}}. \tag{8.2}$$

The exponent Q is the leading power-law analogue to $\hat{\mathsf{Q}}$ of Eq.(2.52), which is well established at least for PBC's. For the jth zero and the magnetic susceptibility it gives

$$h_j(L) \sim \left(\frac{j}{L^d}\right)^{\frac{\Delta \mathsf{Q}}{\nu d}} \tag{8.3}$$

and

$$\chi_L \sim L^{\frac{\gamma \digamma}{\nu}}, \qquad (8.4)$$

respectively. Eq.(8.1) then gives

$$L^{\frac{\gamma \digamma}{\nu}} \sim L^{\frac{2\Delta \digamma}{\nu} - d} \sum_{j=1}^{L^d} j^{-\frac{2\Delta \digamma}{\nu d}}. \qquad (8.5)$$

Matching both sides, and using the static scaling relations $2\beta + \gamma = 2 - \alpha$ and $\beta(\delta - 1) = \gamma$, one obtains

$$\frac{\nu d}{\digamma} = 2 - \alpha. \qquad (8.6)$$

This recovers standard hyperscaling (2.37) provided $\digamma = 1$ below d_c. Above d_c, it delivers $\nu d_c = 2 - \alpha$, which is also correct.

The FSS form $\chi_L \sim L^2$ corresponds to $\digamma = 1$. Eqs.(8.1) and (8.3) would then deliver a leading logarithm in the $d = 6$ Ising case. Such a result is spurious; leading logarithms can occur *only* at the upper critical dimension.[9,27,73,74] Indeed, Butera and Pernici have recently given convincing evidence for the absence of leading logarithms in the Ising and scalar-field models using high-temperature series expansions in the thermodynamic limit in six dimensions.[75] This indicates that $\digamma \neq 1$ *even for FBC's*.

These conclusions are backed up by numerical simulations. For the PBC case, however, either of the above paradigms supports the scaling forms $m_L \sim L^{-d/4}$ and $\chi_L \sim L^{d/2}$, so these cannot be used to discriminate between the two scaling scenarios. Direct numerical confirmation of Eq.(8.2) in the PBC case, however, was given in Refs.[47,53,54]. The formula (8.3) for the Lee-Yang zeros was also verified for PBC's in Ref. [53].

As discussed above, the long-standing belief is that Landau FSS rather than Q-FSS holds for FBC's above the upper critical dimension, at least at the critical point.[60–63] In Refs. [53,54], however, numerical evidence in favour of Q-FSS for FBC's was presented: The claim is that Q-FSS applies in the FBC case above $d = 4$, just as it does in the PBC case, but to observe it, one must perform FSS at the pseudocritical point for FBC's, not at the critical one. The claim is that the infinite-volume critical point is so far from the pseudocritical point as to be outside the Q-FSS window.

For a size-L hypercubic FBC lattice the boundary sites interact with fewer than $2d$ nearest neighbours. In this sense, they belong to a manifold of lower dimensionality. To truly probe the dimensionality of the

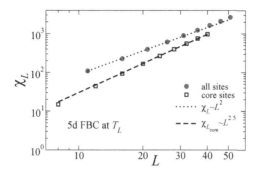

Figure 1. FSS for the 5D Ising susceptibility at pseudocriticality. For the upper data set (red) all sites are used in calculating χ_L. For the lower set (blue) only core sites contribute. While the upper data set may appear to scale as $\chi_L \sim L^2$, in accordance with Gaussian FSS, this is spurious and due to surface sites. The scaling of the lower data, which are genuinely five-dimensional, supports the Q-FSS form $\chi_L \sim L^{5/2}$.

FBC lattices, the contributions of the outer layers of sites can be removed from the susceptibility and from other observables. To implement this in Refs. [53,54], the contributions of the $L/4$ sites near each boundary were removed and only the contributions of the $(L/2)^d$ interior sites kept. For technical details of the simulation and determination of the core thermodynamic functions, the reader is referred to Refs. [53,54]. The resulting FSS for the susceptibility at pseudocriticality for $d = 5$ is represented in Fig. 1.

The upper data set corresponds to using all sites in the calculation of the susceptibility. The fit to the form $\chi_L \sim L^2$ (dotted line), suggested by the traditional Gaussian FSS paradigm for FBC lattices, appears reasonable at first sight. However, closer inspection shows some deviation of the large-L data from the line. We interpret this as signaling that the apparent fit to $\chi_L \sim L^2$ is spurious. The lower data set corresponds using only the interior $L/2$ lattice sites in the calculation of χ_L. The best fit to the Q-FSS form $\chi_L \sim L^{\koppa\gamma/\nu} = L^{5/2}$ (dashed line) describes the large-L data well. This is evidence that the Ising model defined on the five-dimensional core of the L^5 lattices obeys Q-FSS (7.1) at pseudocriticality rather than Gaussian or Landau FSS. This can be interpreted as evidence for the universality of $\koppa$.

The susceptibility is closely related to the Lee-Yang zeros, as we have seen. The FSS for the first two Lee-Yang zeros is presented in Fig. 2 at the pseudocritical point for FBC's using the contributions from all sites and from the core-lattice sites only. In each case the zeros scale as $L^{-\koppa\Delta/\nu} = L^{-15/4}$ according to Q-FSS. There is no evidence to support the Gaussian prediction that $h_j(L) \sim L^{-\Delta/\nu} = L^{-3}$.

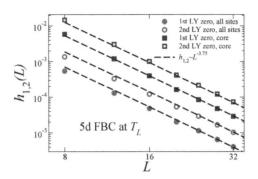

Figure 2. The first two Lee-Yang zeros for Ising systems with FBC's at pseudocriticality obey Q-FSS whether the full lattice or only the core sites are used in their determination.

Thus we arrive at a new *universal* scaling picture, to replace the old non-universal one. The FSS ansatz is Eq.(7.1) with correlation length given by Eq.(8.2). When $d > d_c$, $\mathvarsigma = d/d_c$ and when $d < d_c$, $\mathvarsigma = 1$ and the FSS ansatz reverts to Eq.(2.27). There is no requirement for the thermodynamic length in the scaling ansatz. Also, hyperscaling is extended through Eq.(8.6) beyond the upper critical dimension. The new exponent $\mathvarsigma$ is therefore both *physical* and *universal* in the new picture. Physically it controls the finite-size correlation length. We refer to this picture as Q-theory or Q-FSS, to distinguish it from Gaussian or Landau FSS.

Thus we conclude that the numerical evidence is in favour of Q-FSS in the FBC case at pseudocriticality, just as it is for PBC's. It turns out that at the critical point itself, however, Q-FSS is not supported even when one examines the core 5D FBC lattices. The reason for this is that T_c lies outside the Q-scaling window which is measured by the rounding of the susceptibility peak.[60] In other words, the shifting $T_L - T_c$ exceeds the rounding. The reader is refered to Refs. [53,54] where numerical evidence supporting this interpretation is given.

This now brings us to re-examine the correlation function.

8.2. *Fisher's Scaling Relation and the Exponent η_Q*

In Ref. [70] a new explanation for the negativity of the measured value of the anomalous dimension was proposed, based on the Q-theory outlined here.[53,54] According to the Q-FSS, there is a difference between the underlying length scale L of the system above d_c and its correlation length ξ_L. In Eq.(2.23), the distance r is implicitly measured on the correlation

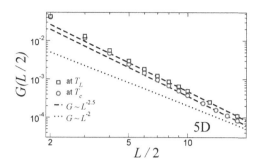

Figure 3. The measured decay of the correlation function in five dimensions favours the Q-prediction $G_Q(t, L/2) \sim L^{-5/2}$ (dashed lines), corresponding to a negative value $\eta_Q = -1/2$, over the Landau prediction of a vanishing anomalous dimension (dotted line).

length scale and this leads to the usual Fisher relation (2.40) via the process outlined in Section 3. In Sec. 7.1, however, the length scales L and ξ_L are incorrectly mixed above the upper critical dimension.

To repair this, we firstly note that Eq.(8.2) with $\digamma = d/d_c$ may be written

$$\xi_L^{d_c} = L^d. \tag{8.7}$$

Thus the correlation volume matches the actual volume when the correct dimensionalities are used, and that for ξ_L is d_c rather than d.

We write the critical correlation function in terms of the system-length scale as

$$G_Q(0, r) \sim r^{-(d-2+\eta_Q)} D_Q\left(\frac{r}{L}\right), \tag{8.8}$$

where η_Q is the anomalous dimension measured on this scale. The subscript reminds that Q-FSS rather than standard FSS prevails there.

Integrating over space, then, the susceptibility is

$$\chi_L(0) \sim \int_0^L r^{1-\eta_Q} D_Q\left(\frac{r}{L}\right) dr = L^{2-\eta_Q} \int_0^1 D_Q(y) y^{1-\eta_Q} dy. \tag{8.9}$$

Above $d = d_c$, the Q-FSS formulae (8.4) then yields

$$\eta_Q = 2 - \frac{\digamma \gamma}{\nu}. \tag{8.10}$$

In the Ising case, for which $\digamma = d/4$, $\gamma = 1$ and $\nu = 1/2$, this gives $\eta_Q = 2 - d/2$. This is then identified with η^* of Refs. [41,43,67,68]. The expression (8.9) with $\eta_Q = -1/2$ in 5D is confirmed in Figs. 3 and 4.

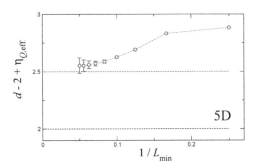

Figure 4.   As the minimum lattice size used in the fit increases, the measured exponent in five dimensions approaches the Q-prediction $\eta_Q = -1/2$.

There are therefore two forms for the correlation function, two anomalous dimensions and two Fisher relations, depending on whether distance is measured on the scale of L or $\xi_L \sim L^\digamma$. The usual value $\eta = 0$ is correct only when distance is measured on the scale of correlation length. The value $\eta_Q = 2 - d/2$ is correct on the length-scale L. Both are valid as characterising long-distance correlation decay. In this interpretation, the notion of two different decay modes, one for short long distances and one for long long distances, is abandoned. The new interpretation is fully consistent with explicit calculations.[41,43,67,68,70]

The relationship between the two anomalous dimensions is

$$\eta_Q = \digamma\eta + 2(1 - \digamma). \tag{8.11}$$

Below d_c, they coincide, while above d_c, the Q-anomalous dimension η_Q is negative. Since the theorems on the non-negativity of the ϕ^4 anomalous dimension refer to correlation decay on the scale ξ, they refer to η rather than η_Q, and are not violated.[23,25,69]

9. The Full Scaling and FSS Theory above the Upper Critical Dimension

In the previous section we confirmed Q-FSS numerically and demonstrated that it applies both to FBC's and PBC's. We interpret this as evidence for universality of FSS and of the new exponent $\digamma$ above d_c. We now need to revisit the renormalisation group scheme and show how the full scaling and FSS theory arises through the mechanism of dangerous irrelevant variables.

9.1. Leading Scaling Behaviour

We start with Eq.(6.27) for the free energy:

$$f_L(t, h, u) = b^{-d} f_{L/b} \left(t b^{y_t}, h b^{y_h}, u b^{y_u} \right). \tag{9.1}$$

Under the assumption of homogeneity (6.28) we have the more compact form (6.29). Identifying d^* with d using the argumentation of Sec. 6.5, one has $p_1 = 0$ in Eq.(6.30). The free energy is then

$$f_L(t, h, u) = b^{-d} \bar{f}_{L/b} \left(t L^{y_t^*}, h L^{y_h^*} \right), \tag{9.2}$$

in which mean-field values $\alpha = 0$, $\beta = 1/2$, $\gamma = 1$ and $\delta = 1/3$ come from

$$y_t^* = y_t - \frac{y_u}{2} = \frac{d}{2}, \tag{9.3}$$

$$y_h^* = y_h - \frac{y_u}{4} = \frac{3d}{4}. \tag{9.4}$$

Appropriate differentiation of the free energy delivers the correct, Landau scaling in the thermodynamic limit and the correct Q-FSS forms (6.45), (6.46), irrespective of boundary conditions.

In Ref. [36], extensions of the above scenario to the correlation length or correlation function were not fully considered. This is because (a) the finite-size correlation length was believed to be bounded by the system length and (b) the correct Landau values for ν and η were obtained by ignoring u. Now we recognise that (a) is incorrect since $\digamma > 1$ and (b) a new exponent η_Q also emerges above d_c.

The equivalent form to Eq.(6.27) for the correlation length is

$$\xi_L(t, h, u) = b \xi_{L/b} \left(t b^{y_t}, h b^{y_h}, u b^{y_u} \right). \tag{9.5}$$

Following a similar procedure to that used for the free energy, the correlation length is expressed as a homogeneous function in Eq.(6.49). Eq.(6.50) gives $\digamma = 1 + q_1 y_u = d/4$ provided $q_1 = -1/4$. Setting $h = 0$ and $L \to \infty$ recovers $\xi_\infty \sim t^{-\nu}$ provided $q_2 = -1/2$ so that $y_t^{**} = d/2$. Thus y_t^{**} is the same as y_t^* of Eq.(9.3). It is also the same as the scaling dimension required in Section 6.5, reaffirming that FSS is controlled by ξ and the thermodynamic length ℓ is not required. Moreover, $q_2 = p_2$ after Eq.(6.44).

To determine q_3 and y_h^{**}, we set $t = 0$ and let $L \to \infty$ and compare to $\xi_\infty(h, 0)) \sim h^{-1/3}$ after Eqs.(2.21) and (2.42). One finds $q_3 = p_3 = -1/4$ and $y_h^{**} = y_h^* = 3d/4$.

Finally we return to the correlation function. According to the standard paradigm, this is not affected by dangerous irrelevant variables. Q-theory,

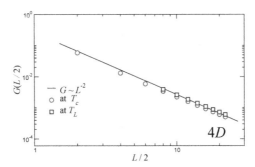

Figure 5.   The leading scaling for the correlation function in four dimensions has exponent $d_c - 2 + \eta = 2$, corresponding to vanishing anomalous dimension. This is predicted by both paradigms.

however, demands that it deliver correlation decay both on the scale of system length as well as on the correlation-length scale.

From dimensional analysis, we write

$$G_L(t, u, r) = b^{-2d_\phi} G_{L/b}\left(tb^{y_t}, ub^{y_u}, rb^{-1}\right), \qquad (9.6)$$

in which $d_\phi = d/2 - 1$. Here, r and b initially have the same dimension as L. Acknowledging the danger of u, we treat this in a similar manner to the free energy and correlation length, and demand

$$G_L(t, u, r) = b^{2-d+y_u v_1} \bar{G}_{L/b}\left(tb^{y_t^*}, rb^{-1+y_u v_2}\right). \qquad (9.7)$$

Maintaining our interpretation of r as a length requires $v_2 = 0$ so that the final argument on the right is dimensionless. Setting $t = 0$ and $b = r$ then, we obtain $G_L(0, u, r) = 1/r^{d-2-y_u v_1}$. This agrees with G_Q in Eq.(8.8) provided that $v_1 = -\eta_Q/y_u = -1/2$.

If, on the other hand, we wish to interpret r as a correlation length, the final argument is dimensionless if it is $rb^{-\digamma}$. We then require $v_2 = (1 - q)/y_u = 1/4$. Again setting $t = 0$, but now with $b^\digamma = r$, we obtain $G_L(0, u, r) = 1/r^{(d-2-y_u v_1)/\digamma}$. Inserting $v_1 = -1/2$ delivers the Ornstein-Zernike[31] form $G(0, u, r) \sim 1/r^2$, corresponding to the Landau theory.

9.2. Logarithmic corrections

Although the theories underlying Q-scaling and the conventional paradigm differ at a fundamental level, each scheme predicts the same numerical values for the various critical exponents in the PBC case. To decide between them, we examine the upper critical dimension itself. The quartic variable

u is marginal there and the renormalization-group formalism gives rise to logarithmic corrections. With no dangerous irrelevant variables, the old paradigm has no consequence at $d = d_c$: there is only one correlation function for long distances. In fact, taking logarithmic corrections into account, one expects at criticality

$$G_\xi(0, r) \sim D\left(\frac{r}{\xi_L}\right) r^{-(d_c - 2 + \eta)} (\ln r)^{\hat{\eta}}. \tag{9.8}$$

The scaling relations for logarithmic corrections connect the logarithmic analogue of the anomalous dimension, $\hat{\eta}$ to η, $\hat{\gamma}$ and $\hat{\nu}$ through Eq.(2.56), which was proposed and confirmed in the thermodynamic limit in Ref. [27] and reported upon in the previous volume of this series.[2]

Q-scaling theory, however, again leads to a new prediction here. The Q-correlation function is

$$G_L(0, r) \sim r^{-(d - 2 + \eta_Q)} (\ln r)^{\hat{\eta}_Q}. \tag{9.9}$$

Of course, $\eta_Q = \eta$ at $d = d_c$.

The fluctuation-dissipation theorem, with an appropriate bound L for Eq.(9.9) then gives[70]

$$\chi_L(0) \sim L^{2 - \eta_Q} (\ln L)^{\hat{\eta}_Q} \left[1 + \mathcal{O}(1/\ln L)\right], \tag{9.10}$$

where

$$\hat{\gamma} = (2 - \eta_Q)(\hat{\nu} - \hat{\varphi}) + \hat{\eta}_Q. \tag{9.11}$$

The two anomalous dimensions are related by

$$\hat{\eta}_Q = \hat{\eta} + (2 - \eta)\hat{\varphi}. \tag{9.12}$$

These new predictions are not derivable from the conventional paradigm and can be tested numerically.[70] In Fig. 5, $G(t, L/2)$ is plotted against $L/2$ at both the critical and pseudocritical points for periodic lattices, confirming the leading behaviour is governed by a vanishing anomalous dimension. The Q-prediction is that $\hat{\eta}_Q = 1/2$ in the $d = 4$ Ising model. This is confirmed in Fig. 6, in which $(L/2)^2 G(t, L/2)$ is plotted against $\ln(L/2)$. The positive slope is clearly not $\hat{\eta} = 0$ and compatibility with $\hat{\eta}_Q = 1/2$ is evident.

10. Conclusions

Mean-field and Landau theories are presented in the early chapters of many textbooks on critical phenomena because (a) they are a simple theories

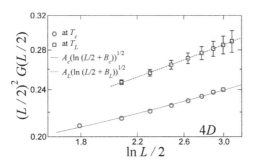

Figure 6.   The logarithmic correction to scaling for the measured correlation function
has a positive exponent compatible with the Q-prediction $\hat{\eta}_Q = 1/2$ and incompatible
with the Landau value zero.

which manifest phase transitions and (b) they form the starting point for
the development of more sophisticated and realistic theories. Despite their
simplicity, the Ginzburg criterion indicated that mean-field and Landau
theories should deliver a comprehensive account of scaling above the upper
critical dimension. In particular, they yield the critical exponents $\alpha = 0$,
$\beta = 1/2$, $\gamma = 1$, $\delta = 1/3$, $\nu = 1/2$ and $\eta = 0$. These satisfy all scaling
relations except hyperscaling (2.37) when $d > 4$.

On the theoretical side, it has long been recognised that, to match the
renormalizaton group to the predictions of mean-field and Landau theories in the thermodynamic limit, one needs to take into account dangerous
irrelevant variables above the upper critical dimension. While this delivers the correct values of the critical exponents, a naive application of FSS
then delivers Gaussian behaviour $m_L \sim L^{-1}$ and $\chi_L \sim L^2$, for example.
These are known to be incorrect for systems with PBC's where, instead,
$m_L \sim L^{-d/4}$ and $\chi_L \sim L^{d/2}$. To remedy this, a new thermodynamic length
scale was introduced to control FSS above d_c. However Gaussian values
were believed to apply in the FBC case at T_c, so that FSS ceases to be
universal above the upper critical dimension there. Moreover, although the
same arguments enter the thermodynamic and correlation functions, only
some of the former were believed to be affected by the danger of the quartic
coupling above d_c.

Numerical simulations indicated another puzzle related to the correlation function. Despite theorems to the contrary, studies indicated a negative
value for the anomalous dimension, while Landau theory predicts that η
vanishes. To explain this, the idea was formed that there is a difference

between short long-range order and long long-range order. Thus the standard paradigm appears to involve ad hoc fixes to a number of puzzles and involves a multiplicity of length scales.

Here we have reported on recent advances to scaling theory above the upper critical dimension. Firstly it is recognised that earlier bounds on the scaling of the correlation length are inapplicable at least at the pseudocritical point. In fact numerical evidence supports non-trivial FSS for ξ_L and an analytical argument requires the introduction of a new exponent (denoted ϙ) to track it.[a] The same numerical argument suggests that the new exponent ϙ is universal as well as physical and this is supported by careful FSS for systems with free boundaries at pseudocriticality.

These considerations also prompted the re-examination of the correlation function above the upper critical dimension. The claim is that its functional form depends upon whether distance is measured on the system-length scale or on the scale of the correlation length. In the latter case, applicable in the thermodynamic limit, the anomalous dimension vanishes and Fisher's scaling relation applies. However, on the scale of the system length, a new anomalous dimension η_Q has to be introduced. In the case of the Ising model, this is negative and brings with it a new Fisher-type scaling relation. Under the new scaling paradigm, there is no thermodynamic length and no distinction between short long distances and long long distances. These were ad hoc features of the old scaling picture.

The new scaling theory also predicts new results at the upper critical dimension itself. This is the case despite the absence of dangerous irrelevant variables there. In particular, new logarithmic corrections to the correlation function are predicted and tested. The results support the new Q-scaling theory.

Revisiting the Ginzburg criterion, in Sec. 4.4, we see that the replacement of the finite-size volume N by ξ_∞^d leading to the thermodynamic-limit relation (4.47) is unjustified. The dimensionality associated with the correlation length is d_c, not d, according to Eq.(8.7). The expression (4.47) should rather read $\chi_\infty/\beta \ll \xi_\infty^{d_c} m_\infty^2$. The Q-corrected Ginzburg criterion then reads $\nu d_c > 2 - \alpha$, which is not satisfied as an inequality. This means the Ginzburg criterion is invalid, even above d_c, so that strictly speaking, mean-field theory is not a full description of scaling there.

---

[a]The exponent ϙ differs from the usual critical exponents α, β, ... in that it applies to finite-size systems at pseudocriticality rather than to infinite-size systems at criticality. In this sense one may wish to refer to it as a "pseudocritical exponent" rather than a critical one.

Finally, and as stated in Ref. [70], the mechanism proposed under the new scaling picture delivers an *emergent* 4-dimensional field theory on the scale of the correlation length. In other words, starting with a generic field theory in $d > d_c = 4$ dimensions on the scale of L, one emerges at the pseudocritical point with a field theory in d_c dimensions on the correlation-length scale. The volume of space, or space-time in quantum field theory, is $L^d = \xi_L^{d_c}$ after Eq.(8.7), which is universal at pseudocriticality, and any attempt to measure dimensionality on the scale of ξ_L delivers d_c instead of d there. Since this is the physical scale, and since $d_c = 4$ in the field theories of the standard model, our universal mechanism may offer a basis for understanding the effective four-dimensional nature of the physical universe.

Note: Since completion of this review, Lundow and Markström (arXiv:1408.5509) simulated 5D systems with free boundaries up to $L = 160$ and their findings at T_c support standard Gaussian FSS there. Our work[53,54,70] also prompted Wittmann and Young to revisit Ising systems above d_c (arXiv:1410.5296). Besides verifying our claim that $\chi_L \sim L^{d/2}$ at T_L, and hence universality of FSS there, they extend consideration of the Fourier modes in PBC's[41] to the case of FBC's,[60] explaining why $\chi_L \sim L^2$ at $T = T_c$ there. Finally, a quantum gravity model utilising some of the new concepts reviewed here was more recently proposed by Trugenberger (arXiv:1501.01408).

References

1. K. Binder, E. Luijten, M. Müller, N.B. Wilding and H.W.J. Blöte, *Monte Carlo investigations of phase transitions: status and perspectives*, Physica A **281**, 112-128 (2000).
2. R. Kenna, *Universal scaling relations for logarithmic-correction exponents*, in Order, disorder and criticality, Advanced problems of phase transition theory Vol.3, Yu. Holovatch (Ed.), World Scientific (Singapore, 2013).
3. E. Ising, *Beitrag zur Theorie des Ferromagnetismus*, Z. Phys. **31**, 253-258 (1925).
4. W. Lenz, *Beiträge zum Verständnis der magnetischen Eigenschaften in festen Körper*, Z. Phys. **21**, 613-615 (1920).
5. C.N Yang, and T.D Lee, *Statistical theory of equations of state and phase transitions I. Theory of condensation*, Phys. Rev. **87**, 404-409 (1952); T.D Lee and C.N. Yang, *Statistical theory of equations of state and phase transitions II. Lattice gas and Ising model*, ibid. 410-419.
6. M.E. Fisher, *The nature of critical points* in Lecture in Theoretical Physics VIIC, edited by W.E. Brittin (University of Colorado Press, Boulder, 1965), p.1-159.

7. D.A. Kurtze and M.E. Fisher, *Yang-Lee edge singularities at high temperatures*, Phys. Rev. B **20**, 2785-2796 (1979).

8. M.E. Fisher and M.N. Barber, *Scaling theory for finite-size effects in the critical region*, Phys. Rev. Lett. **28**, 1516-1519 (1972).

9. E. Brézin, *An investigation of finite size scaling*, J. Physique **43**, 15-22 (1982).

10. M.E. Fisher, *The theory of critical point singularities*, in Critical Phenomena, Proceedings of the 51st Enrico Fermi Summer School, Varenna, Italy, ed. M.S. Green (Academic Press, New York, 1971), pp. 1-99.

11. M.N. Barber, *Finite-Size Scaling*, in C. Domb and J.L. Levowitz (eds.), Phase Transitions and Critical Phenomena Vol. 8, (Academic Press, New York 1983) pp.145-266.

12. B. Widom, *Surface Tension and Molecular Correlations near the Critical Point* , J. Chem. Phys. **43** 3892-3897 (1965); *Equation of State in the Neighborhood of the Critical Point*, ibid. 3898-3905.

13. A.Z Patashinski and V.L Pokrovskii, *Behavior of an ordering system near the phase transition point*, Soviet Physics JETP. **23**, 292 (1966).

14. R.B. Griffiths, *Thermodynamic Functions for Fluids and Ferromagnets near the Critical Point*, Phys. Rev. **158**, 176 (1967).

15. L.P. Kadanoff, *Scaling laws for Ising models near T_c*, Physics (Long Island City), N. **2**, 263-272 (1966).

16. B.D. Josephson, *Inequality for the specific heat: I. Derivation*, Proc. Phys. Soc. **92**, 269-275 (1967); *Inequality for the specific heat: I. Application to critical phenomena*, ibid. 276-284 (1967).

17. J.W. Essam and M.E. Fisher, *Padé Approximant Studies of the Lattice Gas and Ising Ferromagnet below the Critical Point*, J. Chem. Phys. **38**, 802-812 (1963).

18. G.S. Rushbrooke, *On the Thermodynamics of the Critical Region for the Ising Problem*, J. Chem. Phys. **39**, 842-843 (1963).

19. B. Widom, *Degree of the critical isotherm*, J. Chem. Phys. **41**, 1633-1635 (1964).

20. R.B. Griffiths, *Thermodynamic Inequality Near the Critical Point for Ferromagnets and Fluids*, Phys. Rev. Lett. **14**, 623-624 (1965).

21. R. Abe, *Note on the critical behavior of Ising Ferromagnets*, Prog. Theor. Phys. **38** 72-80 (1967).

22. M. Suzuki, *A Theory on the Critical Behaviour of Ferromagnets*, Prog. Theor. Phys. **38** 289-290 (1967); *A Theory of the Second Order Phase Transitions in Spin Systems. II Complex Magnetic Field*, ibid. 1225 (1967).

23. *Correlation Functions and the Critical Region of Simple Fluids*, M.E. Fisher, J. Math. Phys. **5**, 944 (1964).

24. M.J. Buckingham and J.D. Gunton, *Correlations at the Critical Point of the Ising Model*, Phys. Rev. **178**, 848-853 (1969).

25. M.E. Fisher, *Rigorous Inequalities for Critical-Point Correlation Exponents*, Phys. Rev. **180**, 594, (1969).

26. C. Domb and D.L. Hunter, *On the critical behaviour of ferromagnets* Proc. Phys. Soc. **86**, 1147-1151 (1965).

27. R. Kenna, D.A. Johnston, and W. Janke, *Scaling relations for logarithmic*

corrections, Phys. Rev. Lett. **96** 115701 (2006); Self-consistent scaling theory for logarithmic-correction exponents, ibid. **97** 155702 (2006).

28. P. Curie, Lois expérimentales du magnétisme. Propriétés magnétiques des corps à diverses températures, Ann. Chim. Phys. **5**, 289-405 (1895).

29. P. Weiss, L'hypothse du champ moléculaire et la propriété ferromagnétique, J. Phys. Theor. Appl. **6**, 661-690 (1907).

30. L.D. Landau, Zur Theorie der Phasenumwandlungen II, Phys. Zurn. Sowjetunion **11**, 26-35 (1937) [translation in Collected Papers of L.D. Landau, Pergamon Press, 1965, p. 193 and in D. ter Haar, Men of Physics: L. D. Landau, Vol. 11, Pergamon Press, 1965, p. 61].

31. L.S. Ornstein and F. Zernike, Accidental deviations of density and opalescence at the critical point of a single substance, Proc. Acad. Sci. Amsterdam, **17**, 793-806 (1914).

32. V.L. Ginzburg, Fiz. Twerd. Tela **2**, 2031 (1960) [Sov. Phys. Solid State **2**, 1824 (1961)].

33. K.G. Wilson, Renormalization Group and Critical Phenomena. I. Renormalization Group and the Kadanoff Scaling Picture, Phys. Rev. B **4**, 3174 (1971); Renormalization Group and Critical Phenomena. II. Phase-Spce Cell analysis of Critical Behavior, Scaling Picture, Phys. Rev. B **4**, 3184 (1971); K.G. Wilson, The renormalization group: critical phenomena and the Kondo problem, Rev. Mod. Phys. **47**, 773 (1975).

34. M.E. Fisher, Scaling, universality and renormalization group theory, in Lecture notes in physics **186**, Critical phenomena, ed F.J.W. Hahne, (Springer, Berlin, 1983) pp. 1-139.

35. K. Binder, Critical properties and finite-size effects of the five-dimensional Ising model, Z. Phys. B **61**, 13-23 (1985).

36. K. Binder, M. Nauenberg, V. Privman, and A.P. Young, Finite size tests of hyperscaling, Phys. Rev. B **31**, 1498-1502 (1985).

37. Ch. Rickwardt, P. Nielaba and K. Binder, A finite-size scaling study of the five-dimensional Ising model, Ann. Phys. **3**, 483-493 (1994).

38. K.K. Mon, Finite-size scaling of the 5D Ising model, Europhys. Lett. **34**, 399 (1996).

39. G. Parisi and J.J. Ruiz-Lorenzo, Scaling above the upper critical dimension in Ising models, Phys. Rev. B **54**, R3698-R3702 (1996).

40. E. Luijten and H.W.J. Blöte, Finite-size scaling and universality above the upper critical dimension, Phys. Rev. Lett. **76**, 1557-1561 (1996); ibid. 3662 (erratum).

41. E. Luijten and H.W.J. Blöte, Critical behaviour of spin models with long-range interactions, Phys. Rev. B **56**, 8945-8958 (1997).

42. H.W.J. Blöte and E. Luijten, Universality and the five-dimensional Ising model, Europhys. Lett. **38**, 565-570 (1997).

43. E. Luijten, Interaction range, universality and the upper critical dimension, (Delft University Press, Delft 1997).

44. K. Binder and E. Luijten, Monte Carlo tests of renormalization group predictions for critical phenomena in Ising models, Phys. Rep. **344**, (2001) 179-253.

45. N. Aktekin, Ş. Erkoç, and M. Kalay, The test of the finite-size scaling re-

lations for the five-dimensional Ising model on the Creutz cellular automaton, Int. J. Mod. Phys. C **10**, 1237-1245 (1999).

46. E. Luijten, K. Binder and H.W.J. Blöte, *Finite-size scaling above the upper critical dimension revisited: The case of the five-dimensional Ising model,* Eur. Phys. J. B **9**, 289-297 (1999).

47. J.L. Jones and A.P. Young, *Finite-size scaling of the correlation length above the upper critical dimension in the five-dimensional Ising model,* Phys. Rev. B **71**, 174438 (2005).

48. N. Aktekin and Ş. Erkoç, *The test of the finite-size scaling relations for the six-dimensional Ising model on the Creutz cellular automato,* Physica A **284**, 206-214 (2000).

49. Z. Merdan and R. Erdem, *The finite-size scaling study of the specific heat and the Binder parameter for the six-dimensional Ising model,* Phys. Lett. A **330**, 403-407 (2004).

50. Z. Merdan and M. Bayrih, *The effect of the increase of linear dimensions on exponents obtained by finite-size scaling relations for the six-dimensional Ising model on the Creutz cellular automaton,* Appl. Math. Comput. **167**, (2005) 212-224.

51. N. Aktekin and Ş. Erkoç, *The test of the finite-size scaling relations for the seven-dimensional Ising model on the Creutz cellular automaton,* Physica A **290**, 123-130 (2001).

52. Z. Merdan, A. Duran, D. Atille, G. Mülazimoğlu and A. Günen, *The test of the finite-size scaling relations of the Ising models in seven and eight dimensions on the Creutz cellular automaton,* Physica A **366**, 265-272 (2006).

53. B. Berche, R. Kenna and J.C. Walter, *Hyperscaling above the upper critical dimension,* Nucl. Phys. B **865**, 115-132 (2012).

54. R. Kenna and B. Berche, *A new critical exponent ϙ and its logarithmic counterpart ϙ̂,* Cond. Matter Phys. **16**, 23601 (2013).

55. H.G. Katzgraber, D. Larson, and A.P. Young, *Study of the de Almeida-Thouless line using power-law diluted one-dimensional Ising spin glasses,* Phys. Rev. Lett. **102**, 177205 (2009).

56. Derek Larson, Helmut G. Katzgraber, M. A. Moore, A. P. Young, *Numerical studies of a one-dimensional 3-spin spin-glass model with long-range interactions,* Phys. Rev. B **81**, 064415 (2010).

57. F. Beyer, M. Weigel, and M.A. Moore, *One-dimensional infinite-component vector spin glass with long-range interactions,* Phys. Rev. B **86**, 014431 (2012).

58. K. Binder, *Finite size effects at phase transitions,* in Computational Methodes in Field Theory, Proceedings of the 31th Internationale Universitätswoche für Kern- und Teilchenphysik, Schladming, Austria, 1992 edited by H. Gausterer and C.B. Lang (Springer, Berlin) pp. 59-125.

59. J.G. Brankov, D.M. Danchev and N.S. Tonchev, *Theory of critical phenomena in finite-size systems: scaling and quantum effects,* (World Scientific, Singapore, 2000).

60. J. Rudnick, G. Gaspari and V. Privman, *Effect of boundary conditions on critical behavior of a finite high-dimensional Ising model,* Phys. Rev. B **32**, 7594-7596 (1985).

61. P. H. Lundow and K. Markström, *Non-vanishing boundary effects and quasi-first-order phase transitions in high dimensional Ising models*, Nucl. Phys. B **845**, 120-139 (2011).

62. P.G. Watson, *Surface and size effects in lattice models*, in *Phase Transitions and Critical Phenomena* (eds Domb C. & Green M.S.) Vol. 2 101-159 (Academic Press, London, 1972).

63. J.D. Gunton, *Finite size effects at the critical point*, Phys. Lett. **26**, 406-407 (1968).

64. R. Kenna and C.B. Lang, *Finite-size scaling and the zeroes of the partition function in the ϕ_4^4 model,* Phys. Lett. B **264**, 396-400 (1991); *Renormalization group analysis of finite-size scaling in the ϕ_4^4 model,* Nucl. Phys. B **393**, 461-479 (1993); ibid. **411**, 340 (1994).

65. R. Kenna and C.B. Lang, *Scaling and Density of Lee-Yang Zeroes in the Four-Dimensional Ising Model*, Phys. Rev. E **49**, 5012-5017 (1994).

66. R. Kenna, *Finite Size Scaling for $O(N)$ ϕ^4 Theory at the Upper Critical Dimension,* Nucl. Phys. B **691**, 292 (2004).

67. J.F. Nagle and J.C. Bonner, *Numerical studies of the Ising chain with long-range ferromagnetic interactions,* J. Phys. C: Solid State Phys. **3**, (1970) 352.

68. G.A. Baker, Jr. and G.R. Golner, *Spin-Spin Correlations in an Ising Model for Which Scaling is Exact,* Phys. Rev. Lett. **31**, (1973) 22-25.

69. B. Delamotte, D. Mouhanna and M. Tissier, *Nonperturbative renormalization-group approach to frustrated magnets*, Phys. Rev. B **69**, (2004) 134413 (2004).

70. R. Kenna and B. Berche, *Fisher's scaling relation above the upper critical dimension*, EPL **105**, 26005 (2014).

71. C. Itzykson, R.B. Pearson, and J.B. Zuber, *Distributions of zeros in Ising and gauge models*, Nucl. Phys. B **220**, 415-433 (1983).

72. W. Janke and R. Kenna, *The Strength of First and Second Order Phase Transitions from Partition Function Zeroes*, J. Stat. Phys. **102**, 1211-1227 (2001).

73. F.J. Wegner and E.K. Riedel, *Logarithmic corrections to molecular-field behavior of critical and tricritical systems*, Phys. Rev. B **7**, 248-256 (1973).

74. B. Berche, C. Chatelain, C. Dhall, R. Kenna, R. Low, and J.-C. Walter, *Extended scaling in high dimensions*, JSTAT P11010 (2008).

75. P. Butera and M. Pernici, *Triviality problem and the high-temperature expansions of the higher susceptibilities for the Ising and the scalar field models on four-, five-, and six-dimensional lattices*, Phys. Rev. E **85**, 021105 (2012).

Chapter 2

Monte Carlo Simulation of Critical Casimir Forces

Oleg A. Vasilyev [*]

*Max-Planck-Institut für Intelligente Systeme,
Heisenbergstraße 3, D-70569 Stuttgart, Germany,
IV. Institut für Theoretische Physik, Universität Stuttgart,
Pfaffenwaldring 57, D-70569 Stuttgart, Germany.*

In the vicinity of the second order phase transition point long-range critical fluctuations of the order parameter appear. The second order phase transition in a critical binary mixture in the vicinity of the demixing point belongs to the universality class of the Ising model. The superfluid transition in liquid ^{4}He belongs to the universality class of the XY model. The confinement of long-range fluctuations causes critical Casimir forces acting on confining surfaces or particles immersed in the critical substance. Last decade critical Casimir forces in binary mixtures and liquid helium were studied experimentally. The critical Casimir force in a film of a given thickness scales as a universal scaling function of the ratio of the film thickness to the bulk correlation length divided over the cube of the film thickness. Using Monte Carlo simulations we can compute critical Casimir forces and their scaling functions for lattice Ising and XY models which correspond to experimental results for the binary mixture and liquid helium, respectively. This chapter provides the description of numerical methods for computation of critical Casimir interactions for lattice models for plane-plane, plane-particle, and particle-particle geometries.

Contents

[*] vasilyev@fluids.mpi-stuttgart.mpg.de

1. Introduction

In 1948 Hendric Casimir has predicted the phenomenon of an attraction of two perfectly conducting plates in the vacuum.[1] This attractive force is caused by the suppression of the zero level quantum electromagnetic fluctuations. Now this phenomenon is known as quantum (or electromagnetic) Casimir effect. In a critical media in the vicinity of the second-order phase transition long-ranged fluctuations of the order parameter arise. In 1978 Fisher and de Gennes pointed out that the spatial confinement of such fluctuations produces an effective force acting on confining surfaces.[2] This phenomena may be observed in the critical binary mixture in the vicinity of the demixing point or in the liquid ^{4}He in the vicinity of the superfluid transition point. Due to similarity with the quantum Casimir effect, the appearance of forces caused by the spatial confinement of fluctuations in the critical media is now known as *Critical Casimir* (CC) effect (or thermodynamic Casimir effect).[3,4]

The first experimentally observed qualitative manifestation of the CC effect[5,6] is the aggregation in colloidal suspensions close to the demixing point in the critical binary mixture. In the planar geometry quantitative experimental measurements of CC force in the vicinity of the superfluid transition point have been done for wetting layers of ^{4}He.[7,8] Similar measurements of the CC force for a critical binary mixture wetting layer in the vicinity of the consolute point have been reported in Ref. [9]. In these experiments liquid-substrate and liquid-vapor surfaces of a wetting layer suppress critical fluctuations affecting the equilibrium thickness of the film which has been measured experimentally. Later on, the CC interaction forces (of a femto-Newton order) between a colloidal particle and a flat substrate have been measured experimentally[10,11] by using the total internal reflection microscopy. Dependence of these forces on the wetting properties of the substrate has been studied in Ref. [12]. Recently the critical depletion in colloidal suspension has been studied experimentally.[13,14] Colloidal aggregation caused by CC interaction in microgravity conditions has been described in Ref. [15]. From the experimental point of view CC effect provides a method to tune an interaction between colloidal particles. It is possible to switch on interactions between colloids in reversible and controllable way by varying the temperature of a binary liquid mixture. The controlled phase transition of the colloidal suspension in the critical binary mixture has been studied in Ref. [16]. In this study the information about the particle-particle interaction potential has been extracted from the inter-particle pair correlation function.

In accordance with the finite-size scaling theory[17,18] the dependence of the *critical Casimir force* (CCF) f_C (per unit area) on temperature and geometry may be expressed via the CCF universal scaling function ϑ. This function ϑ depends on the shape of confining surfaces, on the universality class of the bulk phase transition, and on *boundary conditions* (BCs), imposed on confining surfaces.[19] For example, in the planar geometry (the film of the thickness L in d-dimensional space) in the vicinity of the transition point the CCF as a function of the temperature T and the film thickness L measured in $k_B T$ units is

$$f_C(T, L) = \frac{k_B T}{L^d} \vartheta \left(\frac{L}{\xi} \right), \tag{1.1}$$

where ϑ is the function of a single scaling variable L/ξ, ξ is the bulk correlation length. For an adequate data analysis of experimental results it is desirable to have a knowledge about properties of the CCF. One can study properties of the CCF scaling function by using appropriately chosen lattice model of the same universality class. The transition in a binary liquid mixture near its consolute point belongs to the bulk universality class of the three-dimensional ($d = 3$) Ising model, and the superfluid transition in liquid ^{4}He belongs to the bulk universality class of the $d = 3$ XY model. Theoretical studies by analytical methods do not provide us a knowledge about CCF scaling function for all sets of geometries and BCs in the full

O. A. Vasilyev

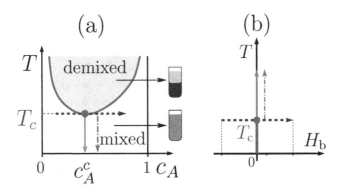

Figure 1.   (a) Phase diagram of a critical binary mixture in the (c_A, T) plane with the lower critical point at (T_c, c_A^c). (b) Phase diagram of the Ising model in the (H_b, T) plane.

range of scaling variables. Using *Monte Carlo* (MC) methods one can compute CCF for relevant universality classes and extract information about appropriate scaling functions.

In Fig. 1(a) the schematic phase diagram of the binary mixture with the lower critical point (e.g., water-lutidine mixture which was used in experiments[5,6,10–12]) is shown. The binary mixture consists of A and B components with concentrations c_A and $c_B = 1 - c_A$, respectively. The critical concentration is c_A^c ($c_B^c = 1 - c_A^c$) and the critical temperature is T_c. The state of this system is characterized by the reduced temperature $t_{AB} = (T - T_c)/T_c$ and chemical potentials μ_A, μ_B for two components A,B with corresponding values μ_A^c, μ_B^c at criticality. The combination of chemical potentials $H_{AB} = \mu_A - \mu_A^c - (\mu_B - \mu_B^c)$ plays a role of the bulk ordering field and another combination $\delta\mu = \mu_A + \mu_B - (\mu_A^c + \mu_B^c)$ describes the deviation of the chemical potential for both components from the critical values. In the most general case, two scaling fields which are linear combinations of these three variables t_{AB} and H_{AB} and $\delta\mu$ characterize the state of the binary liquid mixture in the vicinity of the critical point (see Ref. [20] for detailed description). In Fig. 1(b) the corresponding bulk phase diagram of the Ising model is plotted in the (H_b, T) plane. The state of the model is characterized by the reduced temperature $t = (T - T_c)/T_c$ and the bulk ordering magnetic field H_b. Considering the mapping of parameters of the binary mixture on the Ising model we take equal reduced temperatures $t_{AB} = t$ and the potential difference proportional to the bulk field $H_{AB} \propto H_b$. In Fig. 1(a) we plot critical $c_A = c_A^c$ (solid arrow) and off-critical $c_A >$

c_A^c (dash-dotted arrow) iso-concentration lines and the critical isotherm (dashed arrow). Appropriate thermodynamic paths for the Ising model are reproduced in Fig. 1(b). To define boundary conditions one can translate wetting properties of confining surfaces for a binary mixture to surface fields for the Ising model.

All before-mentioned experiments correspond to the case of $d = 3$ spatial dimensions. Recently it has been shown that in a lipid membrane consisting of two different types of lipids the second order phase transition may occur.[21] Long-ranged fluctuations of the order parameter (e.g., the concentration of lipids of the first type) may cause forces acting on inclusion membrane proteins. This transition belongs to the universality class of the $d = 2$ Ising model. The analytical solution for the $d = 2$ Ising model in rectangular geometry is known for a long time.[22,23] Nevertheless, at the moment only MC simulations can provide rigorous results for disc-disc or disc-wall geometries.

In the case of the superfluid transition in liquid ^{4}He the order parameter is the $n = 2$ component vector (representing the phase of the wave function for the boson condensate) which vanishes at confining surfaces. This experimentally relevant case corresponds to the $d = 3$ XY model without bulk ordering field with open (or Dirichlet) BCs.

The universality of the Casimir force lets us to study its scaling properties in the experimentally relevant cases using representative lattice models. Due to the complexity of the problem analytical studies are limited. Therefore Monte Carlo simulations provide a necessary tool for the computation of the Casimir interactions for various geometries within the whole range of model parameters.

The critical amplitudes of the Casimir force for $d = 2, 3$ Ising model have been studied using hybrid algorithm.[24] The lattice is split into two decoupled subsystems which lets to compute the free energy difference. The scaling function for the $d = 3$ Ising model with *periodic boundary conditions* (PBC) has been computed numerically via the lattice stress tensor.[25] Unfortunately, this approach is suitable only for PBC. The Casimir force as a function of the temperature has been computed using the coupling parameter approach for various types of BCs.[26–28] The same approach has been used for the computation of the force as a function of the bulk ordering field.[29] CCF for a patterned system in plane-plane geometry for Blume-Capel model (Ising universality class) has been studied using coupling parameter approach in Refs. [30,31]. One of boundary planes of the patterned system consists of strips of different wetting preferences, the second plane is homogeneous. Recently, the lines contribution from contacts of stripes of different surface properties to the interaction between homogeneous and patterned planes has been studied.[32] Numerical approach, based on the energy integration over the inverse temperature, has been used for the computation of the Casimir force for the XY model.[33] The same method has been used to study the aspect ratio dependence of CCF for the Ising model with PBC.[34] The energy integration method has been used for the computation of CCF scaling functions for the improved XY model,[35,36] and the improved Ising (Blume-Capel) model with symmetry breaking boundary conditions[37,38] and for studying of the crossover between the ordinary and the normal surface universality classes.[39] Another method for the planar geometry is based on the addition of isolated spins.[40] Recently, the integration over the magnetization has been proposed for the computation of the force for the system with the bulk ordering field.[41] In all abovementioned articles the planar geometry with two boundary surfaces (or PBC) has been considered.

The Casimir interaction between a spherical particle and a plate has been studied by using a version of the energy integration for two systems with joint updates and improved estimator.[42] The CC forces acting on a cubical and spherical particle between two planar walls have been computed using the coupling parameter approach.[43]

The method of the integration over the local field has been used for the computation of the CCF between two spherical particles in the presence of the bulk ordering field.[44] Models, described in these articles belong to the Ising universality class in $d = 3$ spatial dimensions.

Critical interactions between disc-like inclusions in a critical $d = 2$ Ising layer has been studied by the energy integration.[45] A torque, acting on a needle near the wall in a critical system, has been computed using the coupling parameter approach.[46] The alternative method based on the distribution function of a mobile object position has been recently used for computation of the interaction potential between a disc and a wall for $d = 2$ geometry.[47] The phase diagram of ternary solvent-solvent-colloid mixture represented by $d = 2$ Ising model with disc-like particles has been investigated by using grand-canonical insertion technique together with transition matrix MC.[48] In this chapter we discuss Monte Carlo methods for computation of CC interactions. Nevertheless one should mention numerical method based on the minimization of the Ginzburg-Landau Hamiltonian which provides results for the *mean-field theory* (MFT) universality class (formulation of the Ginzburg-Landau model for a planar geometry is provided in Appendix A.5). This method has been applied for studying interactions of a spherical particle with a patterned substrate[49,50] and of ellipsoidal particle with a planar wall.[51] The interaction between two spherical particles in the presence of the bulk field has been studied in Ref. [52] for MFT. This approach also has been used for investigation of many-body interaction in plane-particle-particle[53] and particle-particle-particle[54] systems.

Generally speaking, we can distinguish all systems under considerati- on in $d = 2, 3$ spatial dimensions and in plane-plane, plane-particle and particle-particle geometry types. From the technical point of view the com- putation of the free energy difference is possible by means of integration over the temperature, integration over the bulk field, and coupling param- eter approach (integration over the interaction constant). The application of these methods to various geometries is described in next sections.

2. Casimir forces for planar geometry

2.1. *Model description*

In this section we consider the computation of the CCF in planar (or plane- plane) geometry, which is relevant for the interpretation of experimental results on wetting films of critical liquids.[7–9] We consider Ising and XY models defined on a simple cubic lattice with lattice spacing 1. On the lattice all lengths are measured in units of lattice spacing. For the planar (film) geometry the system has a size $L_x \times L_y \times L_z$ with $L_x = L_y$ and with a cross section $A = L_x \times L_y$. There are periodic boundary conditions along the x and y directions. Each lattice site $i = (1 \leq x \leq L_x, 1 \leq y \leq L_y, 1 \leq z \leq L_z)$ is occupied by a spin $\mathbf{s}_i$. In the Ising model, $\mathbf{s}_i$ has only one component $s_i \in \{+1, -1\}$, whereas in the XY model $\mathbf{s}_i$ is a two-component vector with modulus $|\mathbf{s}_i| = 1$. The Hamiltonian of the Ising model in the most general case with bulk field H_b and surface fields (H_1^- acting on the bottom and H_1^+ acting on the top layers $z = 1, L_z$, respectively) is

$$\mathcal{H} = -J \sum_{\langle ij \rangle} s_i s_j - H_1^- \sum_{\langle \text{bot.} \rangle} s_j - H_1^+ \sum_{\langle \text{top} \rangle} s_j - H_b \sum_k s_k. \qquad (2.1)$$

Here and in the following the energies and fields are measured in units of the spin-spin interaction constant $J = 1$. The sum $\langle ij \rangle$ is taken over all nearest-neighbor pairs of sites on the lattice and the sum over k runs over all spins. The sum $\langle \text{bot.} \rangle$ is taken over the bottom layer $z = 1$ and sum $\langle \text{top} \rangle$ is taken over the top layer $z = L_z$. This provides the bound- ary conditions on which CCF strongly depends. One of the two species of the binary mixture is preferentially adsorbed at a confining wall which within the Ising model corresponds to the presence of a strong positive (or negative) surface field, denoted as $(+)$ (or $(-)$) BCs. If the surface is neutral with respect to both species one is lead to Dirichlet BCs (denoted as (O)). Dirichlet and $(\pm)$ BCs are the renormalization-group fixed-point boundary conditions corresponding to the so-called ordinary surface universality class (O) and the normal transition surface universality class, respectively.[19,55,56] Experiments[9] on wetting films of classical binary liquid mixtures are in agreement with $(+, -)$ BCs corresponding to a strong opposing preferential adsorption of the two species of the mixture at confining surfaces. Measurements for colloid and the substrate surface preferentially adsorb the same species of the binary mixture (both are hydrophobic or both are hydrophilic) are consistent with $(++)$ BCs of the Ising model (that is equivalent to $(--)$ due to the symmetry with respect to the spin sign change $s_i \rightarrow -s_i$). Situation in which the particle and the surface preferentially adsorb different species of the mixture agrees with $(-+)$ or $(+-)$ BCs of the Ising model.[11,12] The four common types of BCs correspond to $(H_1^-, H_1^+) = (+\infty, +\infty) \equiv (+, +)$, $(-\infty, +\infty) \equiv (-, +)$ (normal transition surface universality class), $(0, +\infty) \equiv (O, +)$, and $(0, 0) \equiv (O, O)$ (Dirichlet-Dirichlet surface universality class).[19] We also consider periodic BCs (layers $z = 1$ and $z = L_z$ are connected). For the XY model the Hamiltonian is $\mathcal{H} = -J \sum_{\langle i,j \rangle} \mathbf{s}_i \cdot \mathbf{s}_j$, and we consider only periodic (PBC) and (O, O) boundary conditions in z direction. For the experimentally relevant case pure ^{4}He one has symmetric (O, O) BCs because the quantum wave function of the superfluid state vanishes at confining interfaces.

For large cross section area $A = L_x \times L_y$, the total free energy $F(\beta, H_1^+, H_1^-, H_b, L_z, A)$ of the film of thickness L_z as a function of the inverse temperature $\beta = 1/(k_B T)$, two surface fields H_1^+, H_1^-, and the bulk magnetic field H_b can be written as

$$F(\beta, H_1^+, H_1^-, H_b, L_z, A) = A[L_z f^b(\beta, H_b) + \beta^{-1} f^{ex}(\beta, H_1^+, H_1^-, H_b, L_z)],$$
(2.2)

were the bulk free energy density per spin (in the limit of the infinitely large system) $f^b(\beta, H_b)$ depends only on β and H_b. The excess free energy per unit area f^{ex} contains two L_z-independent surface contributions in addition to the finite-size contribution $f^{ex}(\beta, H_1^+, H_1^-, H_b, L_z) - f^{ex}(\beta, H_1^+, H_1^-, H_b, \infty)$. The critical Casimir force f_C per unit area A measured in units of $k_B T$ equals the derivative of the excess free energy f^{ex}

with respect to the system thickness L_z

$$f_C(\beta, H_1^+, H_1^-, H_b, L_z) \equiv -\partial f^{ex}(\beta, H_1^+, H_1^-, H_b, L_z)/\partial L_z. \tag{2.3}$$

For the lattice model the derivative in Eq. (2.3) is replaced by a finite difference and $f_C(\beta, L)$ is given by

$$f_C(\beta, H_1^+, H_1^-, H_b, L, A) \equiv -\frac{\beta \Delta F(\beta, H_1^+, H_1^-, H_b, L, A)}{A} + \beta f^b(\beta, H_b), \tag{2.4}$$

where the free energy difference $\Delta F(\beta, H_1^+, H_1^-, H_b, L, A) = F(\beta, H_1^+, H_1^-, H_b, L + \frac{1}{2}, A) - F(\beta, H_1^+, H_1^-, H_b, L - \frac{1}{2}, A)$. In these two expressions the thickness $L \equiv L_z - \frac{1}{2}$ which is attributed to the force is half-integer, because it is expressed via difference for slabs of thickness L_z and $L_z - 1$. Later on we shall denote by half-integer $L = L_z - \frac{1}{2}$ the variable on which the critical Casimir force depends and by the integer variable L_z the thickness of the system for which we perform computations.

2.2. *Coupling parameter approach*

The standard MC methods are not suitable for the computation of quantities, such as the free energy, which can not be expressed as an average over ensemble configurations. Therefore, we compute the free energy difference $\Delta F(\beta, H_1^+, H_1^-, H_b, L, A)$ by using the "coupling parameter approach" (see, e.g., Refs. [27,57,58]). This approach is used to compute the difference $F_1 - F_0$ (here we omit arguments of the function like β, H_b, etc.) between free energies $F_i = -\frac{1}{\beta} \ln \sum_C \exp(-\beta \mathcal{H}_i)$, $i = 0, 1$.

The two models that correspond to two free energies F_0 and F_1 are characterized by two Hamiltonians $\mathcal{H}_0$ and $\mathcal{H}_1$. The configuration space C (a total amount of spins) should be the same for both models. To perform this computation, one introduces the crossover Hamiltonian for an "interpolating" system

$$\mathcal{H}_{cr}(\lambda) = (1 - \lambda)\mathcal{H}_0 + \lambda\mathcal{H}_1, \tag{2.5}$$

where $\lambda \in [0, 1]$ is the interpolating parameter. As λ increases from 0 to 1 the crossover Hamiltonian $\mathcal{H}_{cr}(\lambda)$ (a function of the coupling parameter λ) interpolates between $\mathcal{H}_0$ and $\mathcal{H}_1$. Accordingly, the free energy of the crossover system $F_{cr}(\lambda) = -\frac{1}{\beta} \ln \sum_C \exp(-\beta \mathcal{H}_{cr}(\lambda))$ interpolates between F_0 and F_1. The sums $\sum_C$ are taken over all spin configurations C, which are the same for F_0, F_1 and F_{cr}. The difference $F_1 - F_0$ can be expressed as the integral over the coupling parameter $F_1 - F_0 = \int_0^1 F'_{cr}(\lambda)d\lambda$ where

F'_{cr} is the derivative of $F_{cr}(\lambda)$:

$$F'_{cr}(\lambda) = \frac{dF_{cr}(\lambda)}{d\lambda} = \frac{\sum_{\mathcal{C}}(\mathcal{H}_1 - \mathcal{H}_0)e^{-\beta\mathcal{H}_{cr}(\lambda)}}{\sum_{\mathcal{C}} e^{-\beta\mathcal{H}_{cr}(\lambda)}} = \langle\Delta\mathcal{H}\rangle_{cr}(\lambda), \qquad (2.6)$$

which takes the form of the standard ensemble thermal average $\langle\ldots\rangle_{cr}(\lambda)$ of the energy difference $\Delta\mathcal{H} \equiv \mathcal{H}_1 - \mathcal{H}_0$ with respect to crossover Hamiltonian $\mathcal{H}_{cr}$ for a given value of the coupling parameter λ. The energy difference $\langle\Delta\mathcal{H}\rangle_{cr}(\lambda)$ can be efficiently computed via MC simulations of the lattice model with the Hamiltonian $\mathcal{H}_{cr}$. Finally an integral of the energy difference over the coupling parameter provides the difference of free energies

$$F_1 - F_0 = \int_0^1 \langle\Delta\mathcal{H}\rangle_{cr}(\lambda)\, d\lambda. \qquad (2.7)$$

This common method might be used for computation of the CCF in various geometries. The particular choice of interpolating bonds is provided by the selection of initial and final systems (with the same configuration space $\mathcal{C}$) described by Hamiltonians $\mathcal{H}_0$ and $\mathcal{H}_1$, respectively.

For the planar geometry according to Eq. (2.4) one should compute the difference $\Delta F(L)$ between the free energies $F(L_z)$ and $F(L_z - 1)$. We remind, that we denote $L = L_z - \frac{1}{2}$ and omit other parameters $\beta, H_1^+, H_1^-, H_b, A$ which are supposed to be fixed.

In order to apply the coupling parameter approach for the computation of $\Delta F(L)$ for the system with zero bulk magnetic field $H_b = 0$ we identify[26–28] the model $\mathcal{H}_0$ and the appropriate configuration space $\mathcal{C}$ with the model on the lattice $A \times L_z$ of thickness L_z. Therefore $F_0 = F(L_z)$ as depicted in Fig. 2(a). The final system $\mathcal{H}_1$ is identified with the slab of the area A and thickness $L_z - 1$ plus a two-dimensional layer of size A: $F_1 = F(L_z - 1) + F_{2d}$, as shown in Fig. 2(b). Here $F_{2d}(\beta, A)$ is the free energy of the isolated $d = 2$ layer of area A which depends only on the inverse temperature β. We should include this $d = 2$ area into consideration to have the same configuration space $\mathcal{C}$ (number of spins) for initial and final models. This layer can be located at any position along the z-direction $z_0 = 1, 2, \ldots, L_z$ as shown in Fig. 2(c). Then this layer decouples from the rest of the lattice upon passing from $\lambda = 0$ (Fig. 2 (a)) to $\lambda = 1$ (Fig. 2 (b)). In our simulations we use $z_0 = L_z/2$ for even values of L_z and $z_0 = (L_z - 1)/2$ for odd values of L_z. The resulting crossover Hamiltonian $\mathcal{H}_{cr}(\lambda)$ additionally depends on the original position z_0 of the extracted layer. But the result of the integration in Eq. (2.7) does not depend on z_0. The crossover Hamiltonian $\mathcal{H}_{cr}(\lambda) = \mathcal{H}_0 + \lambda\Delta\mathcal{H}$ is characterized by

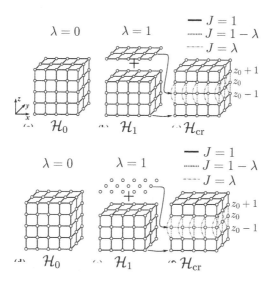

Figure 2. Bond arrangement for the computation of the free energy difference between systems of thickness L_z (a) and $L_z - 1$ (b) in Eq. (2.7). For zero bulk field the crossover Hamiltonian $\mathcal{H}_{\mathrm{cr}}$ (c) belongs to a system which interpolates between those described by the Hamiltonians $\mathcal{H}_0$ (a) for system of thickness L_z (for $\lambda = 0$) and $\mathcal{H}_1$ (b) for system of thickness $L_z - 1$ plus $d = 2$ layer of the area A (for $\lambda = 1$). For nonzero bulk magnetic field Hamiltonian $\mathcal{H}_{\mathrm{cr}}$ (f) interpolates between those described by the Hamiltonians $\mathcal{H}_0$ (d) for system of thickness L_z (for $\lambda = 0$) and $\mathcal{H}_1$ (e) for system of thickness $L_z - 1$ plus A isolated spins.

the coupling constants depicted in Fig. 2(c). Here the dash-dotted lines denote increasing interactions of strength λ while dashed lines denote the decreasing interactions of strength $1 - \lambda$. The energy difference is expressed as

$$\Delta \mathcal{H} = -\sum_{\langle x,y \rangle} \left(s_{x,y,z_0-1} s_{x,y,z_0+1} - s_{x,y,z_0} s_{x,y,z_0-1} - s_{x,y,z_0} s_{x,y,z_0+1} \right), \quad (2.8)$$

where three Cartesian coordinates (x, y, z) identify the lattice site, the sum $\langle x, y \rangle$ is taken over the layer in xy plane, with a coupling strength $J = 1$ (indicated by solid lines in Figs. 2(a) and (b)). In this sum the product of spins, connected by increasing bond (dash-dotted line) is taken with the sign "+" while the product of spins, connected by decreasing bond (dashed line) is taken with the sign "−" (see Fig. 2(c)).

The free energy difference ΔF (see Eqs. (2.4) and (2.7)) can be finally expressed as

$$\Delta F(\beta, H_1^+, H_1^-, L, A) = -\int_0^1 \langle \Delta \mathcal{H} \rangle_{\mathrm{cr}}(\lambda) \mathrm{d}\lambda + F_{2d}(\beta, A) , \qquad (2.9)$$

where the integral is taken for a given values of the inverse temperature β and surface fields H_1^+, H_1^- (here $H_{\mathrm{b}} = 0$). The free energy of $d = 2$ layer $F_{2d}(\beta, A)$ without the bulk magnetic field may be computed in accordance with the analytical expression from [23].

One should slightly modify this computational scheme [29] for the system with nonzero bulk field in order to avoid the numerical computation of the free energy of $d = 2$ layer $F_{2d}(\beta, H_{\mathrm{b}}, A)$ in the presence of the bulk field for which the analytic solution is not known. It would be convenient to take the same initial state $\mathcal{H}_0$ (see Fig. 2(d)) in combination with the final system $\mathcal{H}_1$ as identified with the slab of the area A and thickness $L_z - 1$ plus A isolated spins, see Fig. 2(e). It means, that bonds within the layer should be also removed, as shown in Fig. 2(f) and the explicit expression for the energy difference is

$$\Delta \mathcal{H} = - \sum_{\langle x,y \rangle} \left(s_{x,y,z_0-1} s_{x,y,z_0+1} - s_{x,y,z_0} s_{x,y,z_0-1} - \right.$$
$$\left. - s_{x,y,z_0} s_{x,y,z_0+1} - s_{x,y,z_0} s_{x+1,y,z_0} - s_{x,y,z_0} s_{x,y+1,z_0} \right) . \qquad (2.10)$$

In this case the expression for the free energy difference is computed in accordance with formula

$$\Delta F(\beta, H_1^+, H_1^-, H_{\mathrm{b}}, L, A) = -\int_0^1 \langle \Delta \mathcal{H} \rangle_{\mathrm{cr}}(\lambda) \mathrm{d}\lambda - \frac{A}{\beta} \ln[2 \cosh(\beta H_{\mathrm{b}})],$$
$$(2.11)$$

where the last term is the contribution from A isolated spins in the presence of the bulk field. Once ΔF has been computed, one has still to subtract f^{b} from it (see Eq. (2.4)) in order to obtain the Casimir force for a slab of assigned thickness $L = L_z - 1/2$. But instead of computation of the bulk free energy density $f^{\mathrm{b}}(\beta, H_{\mathrm{b}})$ one can compute the force difference $\Delta f_{\mathrm{C}}(L_1, L_2) = f_{\mathrm{C}}(L_1-1/2) - f_{\mathrm{C}}(L_2-1/2) = -\frac{\beta}{A}[\Delta F(\beta, L_1) - \Delta F(\beta, L_2)]$ for slabs of thicknesses L_1 and L_2. Then it is possible[26,27] to restore the information about the scaling function $\vartheta(L/\xi)$ by using the procedure described in Appendix A.1. Recently this approach has been applied to the computation of the CCF in planar geometry in the presence of the bulk field.[41] This method is convenient if the force depends only on a single variable (e.g, the inverse temperature) because it involves the spline interpolation.

Typically for the computation of the free energy $\Delta F(\beta, H_1^+, H_1^-, H_b, L, A)$ in Eq. (2.4) slabs of thickness $L_z = 10, 15, 20$ with the aspect ratio $L_x/L_z = L_y/L_z = 6$ have been used.[27-29] In order to compute the average $\langle \Delta \mathcal{H} \rangle_{cr}(\lambda)$ one can use hybrid MC method with a mixture of Wolff[59] and Metropolis[60] algorithms. Each hybrid MC step consists (see Ref. [61]) of a flip of a Wolff cluster according to the Wolff algorithm, followed by $3A$ attempts to flip an arbitrary spin and then by $3A$ attempts to flip a spin $s_{x,y,z}$ with $z \in \{z_0 - 1, z_0, z_0 + 1\}$. These attempts are accepted according to the Metropolis rate. For the computation of the thermal average the typical value of MC steps is $5 - 10 \times 10^5$ splitted into 10 series. For every series, one can perform a numerical integration over N_λ points (the value $N_\lambda = 21$ has been used in [27-29]) by using Simpson's integration rule at fixed inverse temperature, surface and bulk fields and the width of the slab L.

2.3. *Energy and magnetization integration*

In the general case one would like to compute the bulk free energy density and obtain the value of CCF in accordance with Eq. (2.4). It has been shown,[28] that it is possible to consider the free energy density for the large system of cubic shape of the size L_{cube}^3, $L_{cube} \geq 128$ with periodic boundary conditions in all directions as the desired bulk free energy density: $f^b(\beta, H_b) \simeq f^{cube}(\beta, H_b, L_{cube} = 128)$ because finite-size corrections to scaling are less than 10^{-6} that is of order of numerical inaccuracy.

The bulk free energy density can be determined via integration of the average energy $E(\beta')$ over the inverse temperature β'

$$\beta f^{cube}(\beta, L_{cube}) = -\ln(2) + L_{cube}^{-3} \int\limits_0^\beta E(\beta') d\beta'. \qquad (2.12)$$

This method has been extended to the case $H_b \neq 0$ in Ref. [29]. In order to obtain f^{cube} for system at the inverse temperature β and with the bulk magnetic field H_b one can integrate the appropriate combination $E(\beta', H_b) - H_b M(\beta', H_b)$ of the energy and the magnetization:

$$\beta f^{cube}(\beta, H_b) = -\ln(2) + L_{cube}^{-3} \int\limits_0^\beta [E(\beta', H_b) - H_b M(\beta', H_b)] d\beta', \qquad (2.13)$$

where $E(\beta, H_b) = -\langle \sum_{\langle i,j \rangle} s_i s_j \rangle$ and $M(\beta, H_b) = \langle \sum_k s_k \rangle$ are thermally averaged energy and magnetization, respectively. Knowing the free energy density $f^b(\beta, H_b^*)$ at a certain value H_b^* of the bulk magnetic field one can compute the bulk free energy density for an arbitrary value of the magnetic field H_b via the integration:

$$\beta f^b(\beta, H_b) = f^b(\beta, H_b^*) - \beta L_{\text{cube}}^{-3} \int_{H_b^*}^{H_b} M(\beta, H_b')dH_b'. \qquad (2.14)$$

Parameters and results of the simulation for the Ising model from Ref. [29] are reproduced below. By using Eqs. (2.13) and (2.14) numerical integrations along the three paths shown in phase diagram Fig. 3(a) have been performed: $(\beta, H_b = 0)$ (solid arrow), $(\beta, H_b = 0.1)$ (dash-dotted arrow), and $(\beta = \beta_c, H_b)$ (dashed arrow). The same paths are plotted in Fig. 3(b) in (H_b, T) plane. We have employed a histogram reweighting technique (see Refs. [61,80], also briefly described in Appendix A.2) for improving the accuracy of the numerical integration. Accordingly, for 256 points of β_i in the interval $[0, 0.3]$ we have computed histograms (averaged over 10^6 MC steps) of the quantities $E(\beta, H_b = 0)$ and $E(\beta, H_b = 0.1) - 0.1M(\beta, H_b = 0.1)$.

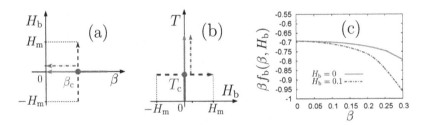

Figure 3. (a) Phase diagram and corresponding paths in the $(\beta = 1/(k_B T), H_b)$ plane: the isotherm runs in the interval $|H_b| \leq H_m$ at β_c (dashed arrow) and two paths $\beta < \beta_c$ for $H_b = 0$ (solid arrow), $H_b = 0.1$ (dash-dotted arrow). (b) Phase diagram of the Ising model in the (H_b, T) plane. (c) The bulk free energy density $f^b(\beta, H_b)$ as a function of the inverse temperature β for $H_b = 0$ (solid line) and $H_b = 0.1$ (dash-dotted line) which correspond to appropriate arrows from the (a) panel.

We have also computed the histogram of $M(\beta_c, H_b)$ for 256 points of the bulk field H_b in the interval $0 \leq H_b \leq H_m$ where $H_m = 0.59$ (see Fig. 3(a)). For negative values of H_b we have used the symmetry relation $f^b(\beta, -H_b) = f^b(\beta, H_b)$. In a second step we have performed numerical integration along these trajectories using histogram reweighting with the trapezoid rule using 10^5 points. Integrating along the solid line $(\beta, H_b = 0)$ we have obtained the critical value $f^b(\beta_c, H_b = 0) = -0.77785038(36)$ whereas sequentially integrating along the dash-dotted $(\beta, H_{\mathrm{b}} = 0.1)$ and dashed $(\beta_c, H_{\mathrm{b}})$ lines, we have obtained the value $f^{\mathrm{b}}(\beta_c, H_{\mathrm{b}} = 0) = -0.77784921(60)$. Both values coincide within the numerical accuracy. It it important, that the bulk free energy should be computed with very high precision, numerical inaccuracy should be less than 10^{-6}. It is required for computation of the CCF scaling functions. The procedure of computation of the CCF consists of the following steps:

- For each value of the inverse temperature β and the bulk field in accordance with Eqs. (2.12),(2.13), and (2.14) we compute 10 estimates for the bulk free energy $f^{\mathrm{b}}(\beta, H_{\mathrm{b}})$ over 10 series of histograms.
- For the given geometry $A \times L_z$, the inverse temperature β, and the boundary H_1^-, H_1^+, and bulk H_{b} fields we compute 10 estimates $\Delta F(\beta, H_1^+, H_1^-, H_{\mathrm{b}}, L, A)$ in accordance with Eqs. (2.9), (2.11).
- Combining the information about the bulk free energy and the free energy difference and using Eq. (2.4) we obtain a numerical estimation for the critical Casimir force $f_{\mathrm{C}}(\beta, H_1^+, H_1^-, H_{\mathrm{b}}, L, A)$ and evaluate the numerical inaccuracy by using averaging over 10 series.

In this section the numerical method for computation of the Casimir force f_{C} for the Ising model is described. The computation of the CCF for the XY model is exactly the same, just instead of the product of the Ising spins $s_i s_j$ the scalar product $\mathbf{s}_i \cdot \mathbf{s}_j$ of $n = 2$ unit vectors $|\mathbf{s}_i| = |\mathbf{s}_j| = 1$ should be used.

2.4. *Finite-size scaling and corrections to scaling*

In accordance with the scaling theory[17,18] the CC interactions are characterized by the ratio of the linear size of the system (the thickness $L = L_z - 1/2$ in the case of the film geometry) and the bulk correlation length $\xi = \xi(t, H_{\mathrm{b}})$ that is the function of the reduced temperature $t = (T - T_c)/T_c = (\beta_c - \beta)/\beta$ and the bulk field H_{b}.

For the $d = 3$ Ising model the value of the critical inverse temperature is $\beta_c = 1/(k_{\mathrm{B}} T_c) = 0.2216544(3)$.[62] In the general case the correlation length is an unknown function of the reduced temperature t and of the bulk field H_{b}. But for zero magnetic field the correlation length is $\xi_t(t) \equiv \xi(t, 0) = \xi_0^{\pm} t^{-\nu}$ and at the critical temperature the correlation length is $\xi_h(H_{\mathrm{b}}) \equiv \xi(0, H_{\mathrm{b}}) = \xi_0^H |H_{\mathrm{b}}|^{-\frac{\nu}{\Delta}}$ where the value of the correlation length critical exponent is $\nu = 0.63002(10)$,[63] bulk magnetic field exponent is $\Delta = 1.5637(14)$,[64] surface gap exponent is $\Delta_1 = 0.46(2)$,[65] and critical amplitudes are $\xi_0^H = 0.278(2)$ (see Appendix A.3, another available value is $\xi_0^H = 0.3048(3)$[66]), $\xi_0^- = 0.243(1)$, and $\xi_0^+ = 0.501(2)$.[62]

For the $d = 3$ XY model the value of the critical inverse temperature is $\beta_c = 0.45420(2)$, the correlation length amplitude is $\xi_0^+ = 0.498(2)$,[67] and the value of the correlation length exponent is $\nu = 0.662(7)$.[64] In this paragraph we also provide numerical values of leading correction-to-scaling exponent in the bulk which takes values $\omega = 0.84(4)$ and $\omega = 0.79(2)$[64] for the three-dimensional Ising and XY universality classes, respectively. Please note, that new more accurate numerical evaluations for some of critical indexes and amplitudes have been presented during the last decade. In this review we provide numerical values, which correspond to cited manuscripts. If different values of these constants have been used in different references, we provide the latest one which is supposed to be more accurate.

On the basis of finite-size scaling,[17,19] in spatial dimension $d = 3$ CCF in units of $k_B T$ and per $d - 1 = 2$-dimensional area are expected to exhibit the scaling form

$$f_C^{(BC)}(\beta, H_b, L) = L^{-3} \vartheta^{(BC)}(L/\xi_t, L/\xi_h), \qquad (2.15)$$

where the ratios $L/\xi_t = \mathrm{sgn}(t) t^\nu L/\xi_0^\pm$, $L/\xi_h = \mathrm{sgn}(H_b) H_b^{\nu/\Delta} L/\xi_0^H$ contain signs of the reduced temperature and magnetic field. It allows one to continue plots for negative values of variables (e.g., $t < 0$ below the critical point $T < T_c$). Instead of these ratios slightly modified scaling variables $H_b \left(L/\xi_0^H\right)^{\Delta/\nu} = (L/\xi_h)^{\Delta/\nu}$ and $t \left(L/\xi_0^+\right)^{1/\nu} = \left(L/\xi_t^+\right)^{1/\nu}$ sometimes have been used, which lets one to avoid the "stretching" of plots in the vicinity of 0. The expression $\xi_t^+ = \xi_0^+ t^{-\nu}$ which includes only high-temperature critical amplitude correlation length ξ_0^+ is the standard way to avoid changing of the slope at zero. The universal scaling function $\vartheta^{(BC)}$ depends on the boundary conditions at the top and at the bottom surface in the case of normal or ordinary surface universality class $(BC) = (+), (-), (O)$. In the case of finite values of surface fields H_1^+, H_1^- (as arguments of f_C) the scaling function ϑ also depends on the scaling variables $H_1^+ L^{\Delta_1/\nu}$, $H_1^- L^{\Delta_1/\nu}$.

The concept of the finite-size scaling assumes, that data points for rescaled force $L^3 f_C$ plotted as a function of the temperature scaling variable $t \left(L/\xi_0^+\right)^{1/\nu}$ (in the case $H_b = 0$) for systems of different thicknesses L should demonstrate the "data collapse", i.e., fall into a single curve. In Fig. 4(a), (b) we plot rescaled force as a function of the scaling variable for system thickness $L = 12.5, 15.5, 19.5$ and for $(+, +)$ and $(+, -)$ BCs, respectively. We observe that curves for different values of L have similar

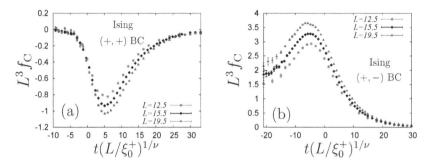

Figure 4. The absence of the data collapse for rescaled Casimir force $L^3 f_C$ for the $d = 3$ Ising model as a function of the scaling variable $t(L/\xi_0^+)^{1/\nu}$ for film thickness $L = 12.5, 15.5, 19.5$: (a) for $(+, +)$ BCs and (b) for $(+, -)$ BCs.

shape but slightly different amplitude. It means, that our system size L is not big enough and we should take into account finite-size corrections to scaling.

Below we collect expressions of corrections to scaling from Refs. [26,27] for the CCF for Ising and XY systems without the bulk field:

$$\text{case (i)}: \quad f_C(\beta, L, A) = L^{-3}(1 + g_1 L^{-1})^{-1}(1 + r_2\rho^2)^{-1}\vartheta\,(x) \ , \quad (2.16)$$

$$\text{case (ii)}: \quad f_C(\beta, L, A) = L^{-3}(1 + g_2 L^{-1})(1 + r_2\rho^2)^{-1}\vartheta(x)\ , \quad (2.17)$$

and

$$\text{case (iii)}: \quad f_C(\beta, L, A) = L^{-3}(1 + g_3 L^{-\omega_{\text{eff}}})\vartheta\,(x) \ , \quad (2.18)$$

where the scaling variable is

$$x = t\left(\frac{L}{\xi_0^+}\right)^{\frac{1}{\nu}}(1 + g_\omega L^{-\omega})(1 + r_1\rho^2), \quad (2.19)$$

where $\rho = L_z/L_x$ is the system aspect ratio, $t = (\beta_c - \beta)/\beta$ is the reduced temperature. The aspect ratio dependent corrections to scaling are relevant only for the XY model with (O, O) BCs.

2.5. Numerical results for the XY model

In Fig. 5(a) (from Ref. [27]) we plot the CCF scaling function ϑ_{OO} for the XY model as a function of x for $L = 9.5, 14.5, 19.5$. Corrections to

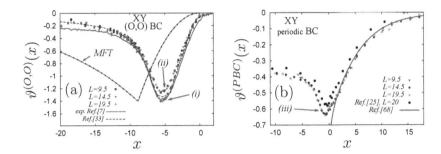

Figure 5. Casimir force scaling function ϑ as a function of the scaling variable x for the three-dimensional XY model with fixed inverse aspect ratio $1/\rho = 6$ and lattice thickness $L = 9.5, 14.5$, and 19.5: (a) CCF scaling function $\vartheta^{(O,O)}$ for (O, O) BCs. The corrections to scaling are taken into account by case (i) and (ii). (b) CCF scaling function $\vartheta_{(PBC)}$ for periodic BC. The corrections to scaling are taken into account by case (iii).

scaling have been accounted for according to two different variants provided by Eq. (2.16) (denoted (i)) and Eq. (2.17) (denoted (ii)). In the first step the value of $L = 10$ has been fixed and data corresponding to different aspect ratios $\rho^{-1} = 4, 5, 6, 8, 10$ have been considered. The values of parameters $r_1 = 1.18(10)$ and $r_2 = 2.40(13)$ are determined using the procedure described in Appendix A.4 such that the data for $(1 + r_2 \rho^2) f_C$ as a function of $t L^{\frac{1}{\nu}} (1 + r_1 \rho^2)$ for fixed L yield the best data collapse onto a single curve. This fitting procedure provides a way to extract corrections to scaling from numerical data for various system sizes (different values of ρ) without knowledge of the fitting function (see also Appendix of Ref. [27] for details). In the second step we fix the value of $\rho = 1/6$ and obtain values of the fitting parameter $g_1 = 5.83(25)$, and $g_\omega = 2.25(15)$ in case (i) and $g_2 = -2.98(8)$, and $g_\omega = 2.25(15)$ in case (ii). With corrections to scaling of the form (ii) the shape of the resulting scaling function is almost the same as for (i) but the amplitude is reduced by a factor $\simeq 0.9$. For (i) our MC data compare very well with the corresponding experimental data from Ref. [7] (solid line) and with the MC data of Ref. [33] (dashed line).

For the XY model with periodic boundary conditions the aspect ratio dependence can not be captured by using r_1, r_2, therefore we perform computations for $\rho = 1/6$. The finite-size corrections have been accounted for case (iii), the resulting value of the exponent is $\omega_{\text{eff}} = 2.59(4)$ in combination with the fitting parameter $g_3 = 14.9(7)$. The CCF scaling function $\vartheta^{(PBC)}$ as a function of the scaling variable x is plotted in Fig. 5(b). The shape of our MC data compares very well with the corresponding MC data (black circles) of Ref. [25]. The solid line corresponds to the analytical prediction in Ref. [68].

2.6. *Numerical results for the Ising model*

For $(+, +)$ and $(+, -)$ BCs the results of simulations of the Ising model do not depend on the aspect ratio $(r_1 = 0, r_2 = 0)$ for $\rho \leq 1/6$. In Figs. 6(a) and 6(b) we plot the scaling function ϑ as a function of the scaling variable $x = t \left(L / \xi_0^+ \right)^{\frac{1}{\nu}} (1 + g_\omega L^{-\omega})$ for $(+, +)$ and $(+, -)$ BCs, respectively. Data points refer to lattices with fixed inverse aspect ratio $1/\rho = 6$ and system thickness $L = 12.5, 15.5, 19.5$. Numerical data are the same as for Fig. 4 where finite-size corrections are not taken into account. Two data sets have been obtained by accounting for corrections to scaling according to Eq. (2.16) (case (i)) and Eq. (2.17) (case (ii)), respectively. For $(+, +)$ BCs the fitting parameters are $g_1 = 14.2(7)$, $g_2 = -4.9(2)$ and $g_\omega = 2.3(2)$ whereas for the $(+, -)$ BCs the fitting parameters are $g_1 = 14.8(2)$, $g_2 = -5.0(1)$, and $g_\omega = 2.9(2)$.[26] In each case the data collapse turns out to be very good. For comparison with $\vartheta^{(+,+)}$ in Fig. 6(a) we provide the prediction of the mean-field theory[69] (solid line), normalized such that $\vartheta^{(+,+)}_{(MFT)}(0) = \vartheta^{(+,+)}_{(MC)}(0)$ (see Appendix A.5), the exact result for the two-dimensional $(d = 2)$ Ising model[70] (dashed line), and the result from the extended de Gennes-Fisher local-functional method[71] (dash-dotted line).

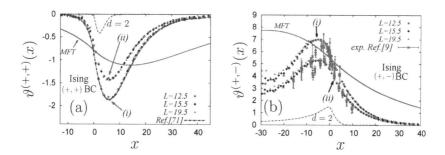

Figure 6. Casimir force scaling function ϑ as a function of the scaling variable $x = t \left(L / \xi_0^+ \right)^{1/\nu} (1 + g_\omega L^{-\omega})$ for the $d = 3$ Ising model with zero bulk field for systems of thickness $L = 12.5, 15.5, 19.5$ and the aspect ratio $\rho = L_z/L_x = 1/6$ for: (a) $(++)$ BCs, (b) $(+-)$ BCs.

To compare the above results with $\vartheta^{(+,-)}$ in Fig. 6(b) we provide the mean-field prediction[69] (solid line), normalized such that $\vartheta^{(+,-)}_{(MFT)}(0) =$

$\vartheta_{(MC)}^{(+,-)}(0)$, the exact result for the two-dimensional Ising model[70] (dashed line), and the set of experimental data points from Ref. [9]. In each case the data collapse turns out to be very good for $x \geq -20$. Within the Derjaguin approximation these numerical results for $\vartheta^{(+,+)}$ and $\vartheta^{(+,-)}$ have been used for calculation of the corresponding scaling functions for the critical Casimir potentials in the plane-sphere geometry. They turn out to be in remarkably good agreement with the actual experimental results.[10] Finite size corrections of form Eqs. (2.16), (2.17), (2.18), and (2.19) have been used in Refs. [26,27,33].

In more recent articles [28,29,35–37,40] another type of finite-size corrections to scaling has been used, which is based on the concept of an *effective* system thickness $L_{\text{eff}} \equiv L + \delta L$ expressed via the correction δL. This effective thickness L_{eff} replaces L in all scaling formulas which in the most general case in presence of boundary and bulk fields read:

$$f_C(\beta, H_1^+, H_1^-, H_b, L) = L_{\text{eff}}^{-3} \vartheta(x, h_1^+, h_1^-, h_b) , \qquad (2.20)$$

where the thermal scaling variable is

$$x \equiv t \left(\frac{L_{\text{eff}}}{\xi_0^+} \right)^{1/\nu} \left(1 + g_\omega L_{\text{eff}}^{-\omega} \right) , \qquad (2.21)$$

the surface field scaling variables are

$$h_1^\pm \equiv H_1^\pm L_{\text{eff}}^{\Delta_1/\nu} , \qquad (2.22)$$

and the bulk field scaling variable is

$$h_b \equiv H_b L_{\text{eff}}^{\Delta/\nu} , \qquad (2.23)$$

while the numerical values of critical exponents are provided in Subsection 2.4. Let us demonstrate the application of these finite-size corrections for a system without the bulk field $H_b = 0$ and with the infinitely strong top surface field $H_1^+ = +\infty$, taking numerical data and figures from Ref. [28]. In the first step by applying the fitting procedure from Appendix A.4 at the critical point $x = 0$ ($\beta = \beta_c$) , we obtain numerical value of the finite-size correction $\delta L = 0.65(2)$. It provides scaling with respect to the surface field scaling variable $\vartheta(0, h_1^-) = L_{\text{eff}}^3 f_C(\beta_c, H_1^-, L)$. In Fig. 7(a) we plot $\vartheta(0, h_1^-)$ as a function of the scaling variable h_1 for the system thickness $L = 9.5, 15.5, 19.5, 24.5$. We observe an excellent data collapse, points for various L fall into the same curve. Then using this value $\delta L = 0.65(2)$ for every selected value of the surface field scaling variable h_1^- we extract information about $g_\omega(h_1^-)$ in accordance with Eq. (2.21) (see Ref. [28] for

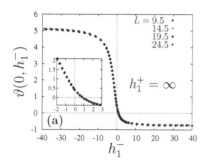 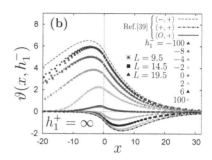

Figure 7. (a) Results for the scaling function $\vartheta(0, h_1^-)$ as a function of the bottom surface field scaling variable h_1^- (Eq. (2.22)) for the finite-size correction $\delta L = 0.65(2)$ at the critical point β_c. (b) The scaling function $\vartheta(x, h_1^-)$ (Eq. (2.20)) as a function of the temperature scaling variable x (Eq. (2.21)) for various values of $h_1^- = -100, -8, -4, -2, 0, 2, 6, 100$.

details). In Fig. 7(b) we plot scaling function $\vartheta(x, h_1^-)$ as a function of the temperature scaling variable $x = t \left(L/\xi_0^+ \right)^{\frac{1}{\nu}} (1 + g_\omega L^{-\omega})$ for various values of $h_1^- = -100, -8, -4, -2, 0, 2, 6, 100$. We also show the MC simulation data obtained for the $d = 3$ Blume-Capel model with $(+, +)$, $(+, -)$ and $(0, +)$ BCs (lines).[39] As we vary the bottom surface field from negative $h_1^- = -100$ to positive $h_1^- = +100$ value, the CC interaction changes the sign from repulsive to attractive.

The same type of finite-size corrections $L_{\text{eff}} = L + \delta L$ has been used for studying of the CCF scaling function as a function of the bulk ordering field H_b exactly at the critical point $(L_{\text{eff}}/\xi_t = 0)$.[29] The CCF has been computed for system thickness $L = 9.5, 14.5, 19.5$ for four types of BSs $(+, +)$, $(-, +)$, $(0, +)$, and $(0, 0)$. From Eq. (2.15) we obtain expression for the CCF scaling function $\vartheta^{(BC)}(L_{\text{eff}}/\xi_h) \equiv \vartheta^{(BC)}(0, L_{\text{eff}}/\xi_h) = L_{\text{eff}}^3 f_C(\beta_c, H_b, L)$. Let us remind, that we include the sign of the bulk field $\text{sgn}(H_b)$ into the ratio $L/\xi_h = \text{sgn}(H_b) H_b^{\nu/\Delta} L/\xi_0^H$. We apply the fitting procedure described in the Appendix A.4 to minimize the spread among data points for various system thicknesses L for each type of BCs. This procedure provides the correction to scaling δL (see Table 1 from Ref. [29]). In this table values of the correction to scaling for $(+, +)$ and $(-, +)$ are in

Table 1. Correction δL to scaling for four BC.

(BC)	$(+, +)$	$(-, +)$	$(0, +)$	$(0, 0)$
δL	0.60(10)	0.65(2)	0.93(10)	1.22(2)

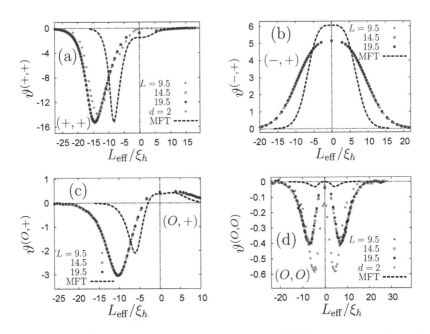

Figure 8.   From Ref. [29]: universal scaling functions $\vartheta^{(BC)}$ as functions of the scaling variable L_{eff}/ξ_h for $d = 3$, $T = T_c$, and $L = 9.5, 14.5, 19.5$, for four BC: (a) $(+,+)$, (b) $(-,+)$, (c) $(O,+)$, and (d) (O,O). The dashed lines show the corresponding normalized MFT results.

a good agreement with the value $\delta L = 0.65(2)$ obtained for a scaling of the surface field (see Fig. 7). In Fig. 8 we plot the results for the CCF scaling function $\vartheta^{(BC)}$ as a function of the scaling variable L_{eff}/ξ_h for the BCs $(+,+)$, $(-,+)$, $(O,+)$, and (O,O), respectively. The value of the critical amplitude $\xi_0^H = 0.278(2)$ (see Appendix A.3) is used for this plot instead of $\xi_0^H = 0.3048(9)$ which has been used in [29]. In this figure we also plot results for the scaling function for *mean-field theory* (MFT) universality class obtained by minimization of the Ginzburg-Landau functional Eq. (A.1) (see Appendix A.5 for details). The undetermined prefactor $g^{-1/2}$ of the MFT has been fixed such that the depths of the minima for functions in Fig. 8(a) is the same. This value of $g \simeq 187.5$ has been used for the MFT results in Figs. 8(b)-(d). For BCs $(+,+)$ and (O,O) (a) and (d) provide a comparison of the scaling functions in $d = 3$ with those in $d = 2$ [72] where one has $\xi_0^H = 0.233(1)$.[73]

3. Casimir forces between a plane and a particle

3.1. *Model description*

In this section a method for computation of CCF between a planar wall and a particle is described.[43] The geometry of a sphere or a cube near a single planar surface is relevant for experiments.[11,12] We use MC simulations of the Ising model on a cubic lattice as a representative of critical phenomena in the Ising universality class. The sketch of the system geometry is shown in Fig. 9(a). A spherical colloidal particle of a diameter a is immersed in critical binary mixture near a planar wall (plane-particle geometry). At the critical point of the demixing transition of the binary solvent, the CCF emerges between the plane and the particle. This force is a result of critical fluctuations of the order parameter of the solvent. Another source of the plane-particle interaction is a critical adsorption of a preferable component of the mixture on the surfaces of the wall and the colloid. The CC force (measured in $k_B T = 1/\beta$ units) between the wall and the particle is the derivative of the free energy $F_C(\beta, D, a)$ of a system consisting of the colloidal particle of the characteristic linear size a with respect to the separation D from the planar wall:

$$f_C(\beta, D, a) = -\frac{\partial(\beta F_C(\beta, D, a))}{\partial D}. \tag{3.1}$$

This definition of the CCF for plane-particle interaction differs from Eqs. (2.3), (2.2) which define the CCF for plane-plane geometry because it does not contain the bulk contribution. We consider the system without the bulk ordering field $H_b = 0$. From the theory of finite-size scaling[17,18] it follows that the CCFs between the wall and the particle can be expressed (neglecting finite-size corrections to scaling) in the following scaling form

$$f_C(\beta, D, a) = \frac{1}{a} K\left(\frac{D}{\xi_t}, \frac{D}{a}\right), \tag{3.2}$$

where the universal scaling function K depends on the ratio of the distance and the bulk correlation length D/ξ_t, and also on the ratio D/a. The force f_C as well as the scaling function K depend on the BCs imposed by the particle and the wall on the order parameter. For the binary mixture this dependence reflects the adsorption preferences of the surfaces of these objects for one of the two species of the mixture.

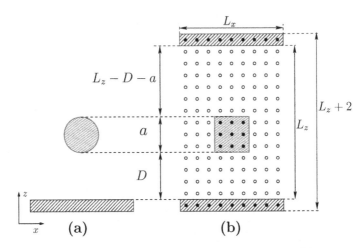

Figure 9.   The geometry of a system for a wall-particle interaction (the cross section in
the xz plane). (a) The spherical particle of the diameter a at the distance D from the
wall. (b) The cubic particle on the lattice with the edge length a at the separation D
from the bottom wall. The total system size is $L_x \times L_y \times (L_z + 2)$, the separation of the
particle from the top wall is $L_z - D - a$, where L_z is the number of layers of free spins.

3.2.   *Coupling parameter approach for wall-particle system*

We consider the Ising model defined on a three-dimensional simple cubic
lattice of the size $L_x \times L_y \times L_z$ via the system Hamiltonian $\mathcal{H}^s$ given by
Eq. (2.1) with periodic BCs in the x and y directions (in which the system
has linear extensions L_x and L_y). There are two additional top and bottom
layers as shown in Fig. 9(b) by filled circles. All spins in the top layer (s_t,
$z = L_z + 1$) are fixed at the value $+1$ whereas all spins in the bottom (s_b,
$z = 0$) layer are fixed at the value $+1$ or -1 for different types of BCs.
We also fix a^3 spins s_p of the cubic particle of linear size a. The spins s_p
can take values $s_p = \pm 1, 0$. The value 0 corresponds to the neutral particle
that does not interact with components of the binary mixture. We have
modified coupling constants between fixed spins at the top and bottom
boundary and those in the adjacent layers as well as between surface of a
particle and their nearest neighbors. The strength of these modified bonds
is infinite, which corresponds to the renormalization group fixed point for
a surface universality class.

   We denote the type of boundary conditions by the expression ($s_b[s_p]s_t$)
which represents values of spins in the (*bottom layer*[*particle*]*top layer*).
For example, $(-[O]+)$ BCs correspond to negative spins $s_b = -1$ at the
bottom layer, zero spins $s_p = 0$ for the particle and positive spins $s_t = +1$

at the top layer. We consider the following cases: a particle between two attractive walls $(+[+]+)$, a particle between the repulsive and attractive walls $(-[+]+)$, and a neutral particle between the walls of opposite signs $(-[O]+)$. For the lattice with unit spacing we rewrite Eq. (3.1) as:

$$f_C(\beta, D - 1/2, a) = -\beta \left[F(\beta, D, a) - F(\beta, D - 1, a) \right] = -\beta \left[F_0 - F_1 \right]. \tag{3.3}$$

The free energy $F(\beta, D, a) = F_0$ corresponds to the system where the particle is at the separation D from the bottom wall (the wall is assumed to be located between the layer at $z = 0$ and the layer at $z = 1$) whereas the free energy $F(\beta, D - 1, a) = F_1$ corresponds to the similar system but with the separation $D - 1$. See Fig. 9 where distances are defined. In order to

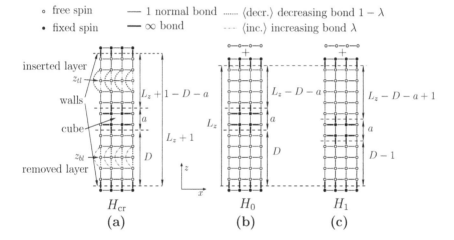

Figure 10. Bond arrangement for the computation of the free energy difference in Eq. (3.7). The crossover Hamiltonian H_{cr} (a) describes the system which interpolates between those described by the Hamiltonians H_0 (b) and H_1 (c).

compute the free energy difference in Eq. (3.3) in accordance with the coupling parameter approach (see Subsection 2.2) we introduce the crossover Hamiltonian $H_{\mathrm{cr}}(\lambda) = (1 - \lambda)H_0 + \lambda H_1$, where $\lambda \in [0, 1]$ is the crossover parameter. This Hamiltonian describes the system consisting of $L_z + 1$ layers of free spins in which the coupling constant associated with certain bonds $\langle \mathrm{inc.} \rangle$ is equal to λ and with some other bonds $\langle \mathrm{decr.} \rangle$ is equal to $1 - \lambda$, as depicted in Fig. 10(a). In this figure, fixed spins of walls and of the particle are denoted by filled circles, while free (fluctuating) spins are denoted by open circles. Bonds of different strength are arranged around two selected horizontal layers of spins. The bottom selected layer is located at $z_{bl} = D - 2$ (at the distance 2 below the particle) and the top selected layer is located at $z_{tl} = D + a + 3$ (at the distance 2 above the particle), as shown in Fig. 10(a). The expression for the crossover Hamiltonian reads

$$H_{\text{cr}}(\lambda) = \tilde{\mathcal{H}}^s - \lambda \sum_{\langle\text{inc.}\rangle} s_i s_j - (1 - \lambda) \sum_{\langle\text{decr.}\rangle} s_i s_j \,, \qquad (3.4)$$

where $\tilde{\mathcal{H}}^s$ is the Hamiltonian of the system with the particle in which bonds around both layers (i.e., at heights $z_{bl} - 1, z_{bl}, z_{bl} + 1$ and $z_{tl} - 1, z_{tl}, z_{tl} + 1$) are removed to avoid the double counting, sums $\langle\text{inc.}\rangle$ and $\langle\text{decr.}\rangle$ are taken over increasing and decreasing bonds. Explicit expressions for sums over increasing bonds (shown by dash-dotted lines in Fig. 10(c)) and decreasing bonds (shown by dashed lines in Fig. 10(c)) in Cartesian coordinates $j = (x, y, z)$ of spins are:

$$\sum_{\langle\text{inc.}\rangle} s_i s_j = \sum_{\langle x,y\rangle} \left[s_{x,y,z_{tl}} \left(s_{x,y,z_{tl}-1} + s_{x,y,z_{tl}+1} \right) + s_{x,y,z_{bl}-1} s_{x,y,z_{bl}+1} \right],$$
$$\qquad (3.5)$$

$$\sum_{\langle\text{decr.}\rangle} s_i s_j = \sum_{\langle x,y\rangle} \left[s_{x,y,z_{bl}} \left(s_{x,y,z_{bl}-1} + s_{x,y,z_{bl}+1} \right) + s_{x,y,z_{tl}-1} s_{x,y,z_{tl}+1} \right],$$
$$\qquad (3.6)$$

where summation $\langle x, y\rangle$ is performed over two-dimensional layers in xy plane, the formula for the energy difference reads

$$\Delta\mathcal{H} = \sum_{\langle\text{decr.}\rangle} s_i s_j - \sum_{\langle\text{inc.}\rangle} s_i s_j \,. \qquad (3.7)$$

For the coupling value $\lambda = 0$, the bottom selected layer at the height z_{bl} is incorporated into the bulk of the system and the top selected layer at the height z_{tl} is excluded from interactions as shown in Fig. 10(b). The configuration described by the Hamiltonian $H_{\text{cr}}(\lambda = 0) = H_0$ corresponds to the system with the free energy F_0 plus the free energy of an isolated 2D layer. The bottom face of the particle is at the distance D (D layers of free spins) from the bottom wall, and the upper face is at the distance $L_z - D - a$ from the top wall. The crossover Hamiltonian $H_{\text{cr}}(\lambda)$ is a function of the coupling parameter λ, it interpolates between H_0 (the system free energy $F_C(\beta, D, a)$) and H_1 (the system free energy $F_C(\beta, D - 1, a)$) as λ increases from 0 to 1. Contributions of 2D layers are mutually canceled, the configuration space is the same. For $\lambda = 1$ the bottom selected layer is excluded from the interactions and the top selected layer is incorporated into the bulk as shown in Fig. 10(c). The total system height L_z remains the same. The free energy of the system described by the Hamiltonian $H_{cr}(\lambda = 1) = H_1$ equals the free energy $F(\beta, D - 1, a)$ plus the free energy of bottom isolated layer. The expression for the Casimir force f_C for the separation $D - 1/2$ between the wall and the cube is

$$f_C(\beta, D_{1/2}, a) = -\beta \int_0^1 \langle \Delta H(\lambda) \rangle_{\{C\}} d\lambda. \tag{3.8}$$

Here, the force as the finite difference between separations D and $D - 1$ is assigned to separation $D_{1/2} \equiv D - 1/2$. Due to the finite size of the system the particle also interacts with the upper wall.

Knowing the critical Casimir force $f_C(\beta, D_{1/2}, a) = -\frac{\partial(\beta F_C)}{\partial D} = \beta(F_C(\beta, D - 1, a) - F_C(\beta, D, a))$, we can compute the potential βF_C as a function of a distance from the wall D with respect to the reference separation D_0. We select as a plane from which the potential is measured the midplane of the system $D_0 = L_z/2$, the potential of the CCFs is zero at this plane. The potential is computed according to the formula

$$\beta F_C(\beta, D, a) = \begin{cases} -\sum\limits_{z=D_0+1}^{D} f_C(\beta, z - 1/2, a), & D \geq D_0, \\ \sum\limits_{z=D_0}^{D-1} f_C(\beta, z - 1/2, a), & D < D_0. \end{cases} \tag{3.9}$$

3.3. *Numerical results for particle-wall interaction*

To demonstrate described above method, the CCF $f_C(\beta, D - 1/2, a)$ and its potential $\beta F_C(\beta, D, a)$ for a cubic particle of edge length $a = 4$ placed in the system of $42 \times 42 \times 45$ free spins have been computed.[43] In Figs. 11(a),(c), and (e) we plot the force f_C and in Fig. 11(b),(d), and (f) its potential βF_C as functions of the distance from the wall. For $(+[+]+)$ BCs the cube is attracted to both walls equally strong. For $(-[+]+)$ BCs the symmetry is broken, the cube is attracted to the $(+)$ wall and it is repelled from the $(-)$ wall. In the case of the "neutral" BCs for cube the potential is symmetric, the particle repels from both walls. The proposed method may be applied not only for particles of the cubic shape, but can be extended to an experimentally relevant plane-sphere geometry.

In the lattice system, we have modeled the sphere of a radius R as follows (this approach for a sphere on the lattice has been used in Ref. [42]). The coordinates of spins on the lattice are integer numbers. We have selected for the center of the sphere the coordinates of the certain central spin. Then we have assigned to the sphere all spins which are located closer to the

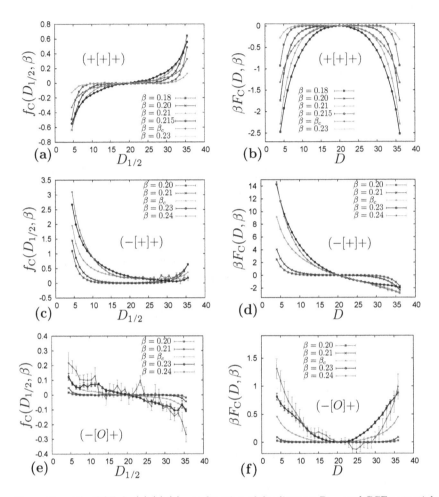

Figure 11. The CCF f_C (a),(c),(e) as a function of the distance $D_{1/2}$ and CCF potential βF_C (b),(d),(f) as a function of the distance D for the cube of size $a = 4$ and various values of the inverse temperature β and for boundary conditions: (a),(b) $(+[+]+)$; (c),(d) $(-[+]+)$; (e),(f) $(-[O]+)$.

central spin than the sphere radius $R = 5.6$. We choose the non-integer radius to provide the right "volume ratio", i.e., the number of spins in the sphere is close to the sphere volume $4/3\pi R^3$. Results for a sphere of radius $R = a/2 = 5.6$ in the system of size $50 \times 50 \times 105$ for $(+[+]+)$ boundary conditions as a function of the inverse temperature β are shown in Fig. 12(a) for various distances $D_{1/2}$ from the bottom wall. The CCFs acting on the particle in the limit of small distance $D \ll R = a/2$ are given

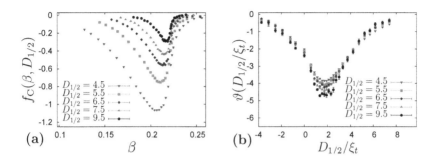

Figure 12.   Results for a sphere of radius $R = 5.6$ placed near the wall with $(+[+]+)$ BCs for different distances $D_{1/2}$: (a) the Casimir force f_C as a function of β, (b) rescaled force $\vartheta = D_{1/2}^2/R$ f_C as a function of $D_{1/2}/\xi_t$.

by the so-called Derjaguin approximation as:[11,74,75]

$$f_C(\beta, D_{1/2}, a) \approx \frac{a}{2D_{1/2}^2}\tilde{K}\left(\frac{D_{1/2}}{\xi_t}\right) , \qquad (3.10)$$

where the scaling function $\tilde{K}(x)$ does not depend on $a = 2R$. At the critical point $\beta = \beta_c$ one has $f_C(\beta_c, D_{1/2}, a) \approx \frac{a}{2D^2}\tilde{K}(0) = \frac{R}{D^2}\tilde{K}(0)$. This observation has encouraged us to apply the "Derjaguin-limit" like rescaling of data obtained for the sphere $R = a/2 = 5.6$ and presented in Fig. 12(b) by the function

$$\vartheta\left(\frac{D_{1/2}}{\xi_t}\right) \equiv \left(\frac{D_{1/2}^2}{R}\right) f_C(\beta, D_{1/2}, 2R). \qquad (3.11)$$

We cannot expect to reach the asymptotic region of the universal behavior Eq. (3.2) in our simulations because of unavoidable limitations of the size of the system and dependence on the variable D/a. Nevertheless for the $(+[+]+)$ case we have obtained acceptable data collapse using Eq. (3.11) even though the size of the sphere R is of order of $D_{1/2}$ and the condition $D_{1/2} \ll R$ is not satisfied.

4. Computation of the critical torque in $d = 2$

4.1. *Model description*

One can consider a lipid membrane (consisting of two or more different types of lipids) as a model system for cell plasma membranes.[76] In such a membrane (which is an $d = 2$ analog of a binary mixture and belongs to the Ising model universality class) the second order phase transition may occur.[77] In the vicinity of the transition point a long-ranged interactions between membrane inclusions arise due to the critical interactions.[21,45,78] Critical interactions between membrane inclusions may be studied by numerical simulation for the $d = 2$ Ising model.

In this section we investigate the critical "torque" that acts on the elongated object ("needle" $[n]$) in a stripe geometry with certain BCs at the bulk critical point of the second order phase transition. It corresponds to plane-particle geometry in $d = 2$. The method and results have been published in Ref. [46]. We shall consider the Ising model with various types of boundaries. We would like to know how the free energy depends on the needle position and orientation. We denote as $F_\parallel$ the free energy of a system with a needle parallel to boundaries and as $F_\perp$ the free energy of a system with a needle perpendicular to boundaries. The "torque" on the lattice is the free energy difference $\Delta F = F_\parallel - F_\perp$ between these two configurations. Detailed definition of the system is given below. Our aim is to investigate the influence of the type of BCs ("normal", "ordinary") and compare numerical results with theoretical evaluations.

The Ising model is simulated on a square lattice of size $L \times W$ where the width of the strip W is substantially smaller than the length L. Spins $s_{x,y} = \pm 1$ are located in sites with coordinates (x, y), with x and y being integers. The coordinate system is defined in such a way, that the point $(x = 0, y = 0)$ corresponds to the middle of the lattice. Our lattice model on the strip is characterized by boundaries (b, t). In order to have the possibility to locate the center of an inserted needle right in the middle between the bottom (b) and the top (t) rows of the lattice we shall use, respectively, an even and odd number W of rows when we consider a needle of broken bonds and of fixed spins introduced in Fig. 13 and Fig. 14.

We define the integer vertical coordinates y by $y^{(b)} \leq y \leq y^{(t)}$ with bottom and top rows at $y^{(b)} = -W/2 + 1$ and $y^{(t)} = W/2$ for W even, and at $y^{(b)} = -W/2 + 1/2$ and $y^{(t)} = W/2 - 1/2$ for W odd. As a consequence the horizontal midline $v = 0$ of the strip is in our lattice models at $y = (y^{(t)} + y^{(j)})/2$ equal to $1/2$ and 0, so that a lattice site with the integer coordinate y has a v-coordinate equal to $v = y - (1/2)$ and $v = y$, respectively. Possible locations v_N for the needle centers are integer values of v, see also Figs. 13 and 14 below. The system is translational invariant in the x direction, the center of the needle always is located on the vertical midline of the system $u = 0$. We consider the system at the bulk critical point $\beta_c = 1/T_c = \frac{1}{2}\ln(\sqrt{2} + 1)$.[22]

We couple the spins in the bottom and top rows with strengths $H_1^{(b)}$ and $H_1^{(t)}$ to spins located in auxiliary rows at $y^{(b)} - 1$ and $y^{(t)} + 1$ and fixed in the $+$ direction in order to implement strip boundaries belonging to the universality classes O, $+$, and $-$. The Hamiltonian of the Ising model $\mathcal{H}^{\mathrm{s}}$ for the strip without the needle reads:

$$\mathcal{H}^{\mathrm{s}} = -J \sum_{\langle \mathrm{nn} \rangle} s_{x,y} s_{x',y'} - H_1^{(b)} \sum_{\langle x \rangle} s_{x,y^{(b)}-1} s_{x,y^{(b)}} - H_1^{(t)} \sum_{\langle x \rangle} s_{x,y^{(t)}} s_{x,y^{(t)}+1},$$

(4.1)

with the interaction constant $J = 1$, the sum $\langle \mathrm{nn} \rangle$ is taken over all pairs $\langle x, y; x', y' \rangle$ of nearest neighbor spins, the sum $\langle x \rangle$ runs over top and bottom layers. Two surface terms couple fluctuating spins of the system with two rows $s_{x,y^{(b)}-1} = 1$ and $s_{x,y^{(t)}+1} = 1$ of fixed spins. The choice of the value of the surface field $H_1^{(b)} = 1$ (or -1) generates a lattice model with a "normal" boundary $(+)$ (or $(-)$) of spins at $y^{(b)} - 1$ that are fixed in the $(+)$ (or $(-)$) direction. Equivalently one may consider it as a boundary at $y^{(b)}$ of spins subjected to a symmetry breaking surface field that points in the $(+)$ (or $(-)$) direction. The choice $H_1^{(b)} = 0$ generates an "ordinary" boundary (O) of spins at $y^{(b)}$ with only three (instead of four) nearest neighbors to which they are coupled. Selecting independently the top surface field $H_1^{(t)} = \pm 1$ or 0 produces a realization of all the pairs (b, t) of BCs for a strip.

The center of the needle of the length D is located in the point with coordinates (u_N, v_N). The insertion of a needle of class $n = O$ or $n = +$ and $n = -$ into the strip requires removing bonds or fixing spins by means of additional terms in the Hamiltonian. Thus we consider a model with the form of a rectangle or "strip" of length L and width W, i.e. $-L/2 < u < L/2$, $-W/2 < v < W/2$ in Cartesian coordinates u, v, and consider a periodic boundary condition along the u-direction while the boundary lines at $v = \pm W/2$ are "ordinary" or "normal" ones. For given boundary classes (b, t) of the strip $\mathcal{H}^{\mathrm{s}}$ Eq. (4.1) and length D, class n, and location v of the needle we adopt the notation

$$\mathcal{H}_0 \equiv \mathcal{H}^{\mathrm{s}} + \mathcal{H}_{\perp}^{[n]}$$

(4.2)

and

$$\mathcal{H}_1 \equiv \mathcal{H}^{\mathrm{s}} + \mathcal{H}_{\parallel}^{[n]}$$

(4.3)

for the complete lattice Hamiltonian $\mathcal{H}$ if the needle is oriented perpendicularly and parallel, respectively, to the $x = u$ axis. The explicit forms of $\mathcal{H}_{\perp}^{[n]}$ and $\mathcal{H}_{\parallel}^{[n]}$ will be given below.

In previous sections we consider the application of the coupling parameter approach for inserting/removing a planar layer of spins. For numerical investigation of critical "torque" we should use the non-planar arrangement of varying bonds. In order to adopt the coupling parameter approach (see Subsection 2.2) we consider two systems with the same configurational space $\mathcal{C}$. For the Hamiltonians $\mathcal{H}_0$ and $\mathcal{H}_1$ given by Eqs. (4.2) and (4.3), respectively, one has

$$\Delta\mathcal{H}^{[n]} = \mathcal{H}_{\parallel}^{[n]} - \mathcal{H}_{\perp}^{[n]} ,$$
$$\mathcal{H}_{\mathrm{cr}}^{[n]}(\lambda) \equiv \tilde{\mathcal{H}}^{\mathrm{s}} + (1-\lambda)\mathcal{H}_{\perp}^{[n]} + \lambda\mathcal{H}_{\parallel}^{[n]} , \qquad (4.4)$$

where $\tilde{\mathcal{H}}^{\mathrm{s}} = \mathcal{H}^{\mathrm{s}} - \mathcal{H}_{\perp}^{[n]}$. As in the previous section, we use notation $(b[n]t)$, $b, n, t = \{O, +, -\}$ to characterize the complete set of bottom b and top t BCs for a needle of a type $[n]$.

4.2. Needle of broken bonds

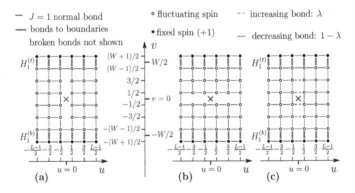

Figure 13. Bond arrangement for a needle $[O]$ of $D = 4$ broken bonds with the center ($\times$) at $(u_N, v_N) = (0,0)$ at the midline $v = 0$ of a strip with $L = W = 6$: (a) for the perpendicular orientation of the needle with Hamiltonian $\mathcal{H}_0$, (b) for the parallel orientation of the needle with Hamiltonian $\mathcal{H}_1$, (c) for the crossover Hamiltonian Eq. (4.4) $\mathcal{H}_{\mathrm{cr}}^{(O)}(\lambda)$ which interpolates between (a) for $\lambda = 0$ and (b) for $\lambda = 1$.

Here we consider the combination $(b[O]t)$ corresponding to an "ordinary" needle which consists of an even number D of broken bonds in a strip with an even number W of rows. In Fig. 13 we show an example of the needle with $D = 4$ broken bonds in the center of an $L \times W = 6 \times 6$ strip. The

Hamiltonians $\mathcal{H}_0$ and $\mathcal{H}_1$ for perpendicular and parallel needle orientation have the same form as the right hand side of Eq. (4.1) but with reduced interaction constants $J_{u,v;u',v'}$ which depend suitably on the coordinates (u, v) and (u', v') of nearest neighbor spins. In Fig. 13 bonds of strength $J = 1$ are indicated by thin lines and broken bonds ($J = 0$) are not shown. The bonds indicated by dash-dotted and dashed lines have strengths λ and $1 - \lambda$ which increase and decrease, respectively, as λ increases. The fluctuating spins $s_{u,v} = \pm 1$ with $-(W - 1)/2 \leq v \leq (W - 1)/2$ are indicated by empty circles while the fixed spins $s_{u,\pm(W+1)/2} = 1$ in the two additional outside layers are indicated by full circles. The nearest neighbor bonds between fixed and fluctuating spins are indicated by thick lines. For the lower and upper boundary b and t they have the strengths $H_1^{(b)}$ and $H_1^{(t)}$, respectively. Surface field H_1 equals 0 for an (O) boundary and to ± 1 for a $(+)/(-)$ boundary. The needle center is located at the point $(0, v_N)$. One can insert the needle in perpendicular orientation by "breaking", i.e., removing the D lattice bonds which at $v = v_N \pm 1/2$, $v_N \pm 3/2$, $\ldots$, $v_N \pm (D-1)/2$ extend from $u = -1/2$ to $u = 1/2$ as shown in Fig. 13(a). This is accomplished by adding

$$\mathcal{H}_\perp^{[O]} = \sum_{\langle \text{inc.} \rangle} \equiv \sum_{k=1}^{D} s_{-1/2, v_N - (D+1)/2 + k} \, s_{1/2, v_N - (D+1)/2 + k} \qquad (4.5)$$

to the Hamiltonian $\mathcal{H}_s$ without the needle (see Eq. (4.2)) where we denote $\sum_{\langle \text{inc.} \rangle}$ the set of increasing bonds. Similarly for the parallel orientation of the needle the sum (which is denoted $\langle \text{decr.} \rangle$)

$$\mathcal{H}_\parallel^{[O]} = \sum_{\langle \text{decr.} \rangle} \equiv \sum_{k=1}^{D} s_{-(D+1)/2 + k, v_N - 1/2} \, s_{-(D+1)/2 + k, v_N + 1/2} \qquad (4.6)$$

should be added to $\mathcal{H}^s$, so that

$$\Delta \mathcal{H}^{[O]} = \mathcal{H}_\parallel^{[O]} - \mathcal{H}_\perp^{[O]} . \qquad (4.7)$$

Fig. 13(a) and 13(b) illustrate these configurations for the special case $v_N = 0$ when the needle is located in the middle of the strip. The sums in Eqs. (4.5) and (4.6) are characterized by subscripts $\langle \text{inc.} \rangle$ and $\langle \text{decr.} \rangle$ because in the crossover Hamiltonian following from Eq. (4.4)

$$\mathcal{H}_{\text{cr}}^{[O]}(\lambda) = \tilde{\mathcal{H}}^s - \lambda \sum_{\langle \text{inc.} \rangle} -(1 - \lambda) \sum_{\langle \text{decr.} \rangle} , \qquad (4.8)$$

they appear with a prefactor $-\lambda$ and $-(1 - \lambda)$, respectively. These sums represent sums of products of spins coupled by nearest neighbor bonds with strengths λ (dash-dotted lines in Fig. 13(c)) and $(1 - \lambda)$ (dashed lines in Fig. 13(c)) which absolute value increase and decrease, respectively, as λ increases. Here $\tilde{\mathcal{H}}^{\mathrm{s}} \equiv \mathcal{H}^{\mathrm{s}} + \sum_{\langle \mathrm{inc.} \rangle} + \sum_{\langle \mathrm{decr.} \rangle}$ equals $\mathcal{H}_{\mathrm{s}}$ in Eq. (4.1) but with both types of nearest neighbor bonds missing which are broken in the perpendicular or the parallel orientation of the needle. This corresponds to Fig. 13(c) with both dashed and dash-dotted bonds removed (so that in $\tilde{\mathcal{H}}^{\mathrm{s}}$ only those bonds of $\mathcal{H}_{\mathrm{s}}$ remain which are outside a hole with the shape of a cross). Therefore in the expression for $\mathcal{H}_{\mathrm{cr}}^{[O]}(\lambda)$ two types of missing bonds are replaced by the bonds of increasing and decreasing strength as illustrated in Fig. 13(c). This leads to the crossover from the perpendicular to the parallel needle orientation as λ increases from 0 to 1.

On this basis, following the steps described by Eqs. ((2.7),(4.5)-(4.7) allows us to calculate $F_{\parallel} - F_{\perp} \equiv F_1 - F_0 = \Delta F$ for the combination of surface fields $(b[O]t)$ for $[O]$ needle.

4.3. Needle of fixed spins

In this subsection we consider the lattice version of the case $(b[+]t)$ in which a needle consisting of an odd number D of spins fixed in the $+$ direction is embedded in a strip with an odd number W of rows. In Figs. 14(a) and 14(b) we show an example of the needle with $D = 5$ in the center of an $L \times W = 8 \times 7$ strip. The strengths H_1 of the bonds near the strip boundaries are the same as in Fig. 13.

The partition functions and free energies of these lattice models remain unchanged if the bonds between the fixed needle spins and their $2D + 2$ fluctuating nearest neighbors are removed. The $2D + 2$ neighbors are coupled instead with the bond interaction constant $J = 1$ to a single exterior spin $s_0 = +1$ as shown in Figs. 15(a) and 15(b). The exterior spin (denoted by a filled circle in top-right corner) is kept fixed in the $+$ direction.

The coupling to this single spin or to the D fixed spins has the same effect on the fluctuating spins. Namely, it acts as a magnetic field acting on the $2D + 2$ neighboring spins. Once one has introduced this coupling to s_0, for the following it is convenient to replace each of the D fixed needle spins by a freely fluctuating spin. This replacement makes contribution $D \ln 2$ to the free energy per $k_{\mathrm{B}}T$. This contribution is independent of the orientation of the needle and thus drops out of ΔF. In Figs. 15(a) and 15(b) additional

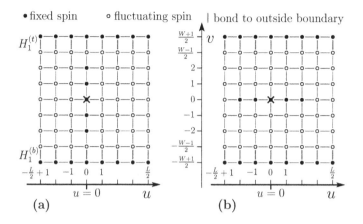

Figure 14. The bond arrangement for a needle of $D = 5$ spins fixed in the $+$ direction (filled circles) with the needle center $(\times)$ located at $(u_N, v_N) = (0,0)$ at the midline $v = 0$ of a strip with $L = 8$ and $W = 7$ for the perpendicular direction of the needle (a) and for the parallel direction of the needle (b).

couplings to the external spin are denoted by top-right arrows.

The corresponding additional terms in the Hamiltonian can be written as

$$\mathcal{H}_{\perp}^{[+]} = \sum_{\langle\text{zero}\rangle} + \sum_{\langle\text{inc.}\rangle} - \overset{(+)}{\sum_{\langle\text{one}\rangle}} - \overset{(+)}{\sum_{\langle\text{decr.}\rangle}}, \qquad (4.9)$$

and

$$\mathcal{H}_{\parallel}^{[+]} = \sum_{\langle\text{zero}\rangle} + \sum_{\langle\text{decr.}\rangle} - \overset{(+)}{\sum_{\langle\text{one}\rangle}} - \overset{(+)}{\sum_{\langle\text{inc.}\rangle}}, \qquad (4.10)$$

so that $\Delta\mathcal{H}^{[+]} = \mathcal{H}_{\parallel}^{[+]} - \mathcal{H}_{\perp}^{[+]}$. Here the sums

$$\sum_{\langle\text{zero}\rangle} = s_{0,v_N}\left(s_{-1,v_N} + s_{1,v_N} + s_{0,v_N+1} + s_{0,v_N-1}\right), \qquad (4.11)$$

$$\sum_{\langle\text{inc.}\rangle} = \sum_{k=1}^{(D-1)/2}\Big[s_{0,v_N+k}\left(s_{-1,v_N+k} + s_{1,v_N+k}\right) + s_{0,v_N-k}\left(s_{-1,v_N-k} + s_{1,v_N-k}\right)+$$

$$+ s_{0,v_N+k}\,s_{0,v_N+k+1} + s_{0,v_N-k}\,s_{0,v_N-k-1}\Big],$$

$$\qquad (4.12)$$

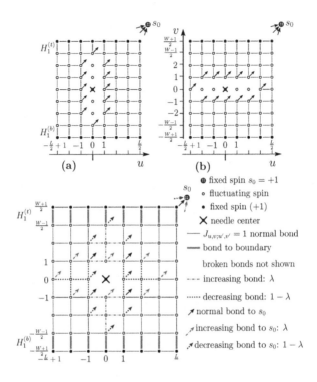

Figure 15. Arrangement of bonds for couplings (top-right arrows) to a fixed external spin $s_0 = +1$ for: (a) perpendicular and (b) parallel orientations of [+] needle (equivalent to configurations in Figs. 14(a) and 14(b), respectively). (c) The arrangement of bonds in the crossover Hamiltonian $\mathcal{H}_{\mathrm{cr}}^{[+]}(\lambda)$ (see Eq. (4.17)) with their strengths indicated in the right margin.

and

$$\sum_{\langle \mathrm{decr.}\rangle} = \sum_{k=1}^{(D-1)/2} \Big[s_{k,v_N}\big(s_{k,v_N+1}+s_{k,v_N-1}\big) + s_{-k,v_N}\big(s_{-k,v_N+1}+s_{-k,v_N-1}\big)$$
$$+ s_{k,v_N}s_{k+1,v_N} + s_{-k,v_N}s_{-k-1,v_N}\Big],$$

$$(4.13)$$

contain products of nearest neighbor lattice spins. The sums

$$\sum_{\langle \mathrm{one}\rangle}^{(+)} = s_{-1,v_N+1} + s_{1,v_N+1} + s_{-1,v_N-1} + s_{1,v_N-1}\,,\qquad (4.14)$$

$$\sum_{\langle \text{decr.} \rangle}^{(+)} = s_{0,v_N+(D-1)/2+1} + s_{0,v_N-(D-1)/2-1} + s_{-1,v_N} + s_{1,v_N}$$

$$+ \sum_{k=2}^{(D-1)/2} \left[s_{-1,v_N+k} + s_{1,v_N+k} + s_{-1,v_N-k} + s_{1,v_N-k} \right], \tag{4.15}$$

and

$$\sum_{\langle \text{inc.} \rangle}^{(+)} = s_{(D-1)/2+1,v_N} + s_{-(D-1)/2-1,v_N} + s_{0,v_N+1} + s_{0,v_N-1}$$

$$+ \sum_{k=2}^{(D-1)/2} \left[s_{k,v_N+1} + s_{k,v_N-1} + s_{-k,v_N+1} + s_{-k,v_N-1} \right], \tag{4.16}$$

contain products of a lattice spin and the fixed external spin s_0 (which has the value $s_0 = +1$ and thus does not appear in Eqs. (4.14)-(4.16)). Here the sum $\sum_{\langle \text{zero} \rangle}$ includes products of the central spin (which is denoted ×) of the needle with its four nearest neighbor spins. These products correspond to lattice bonds broken in both the perpendicular and parallel needle orientation (see Figs. 15(a) and 15(b)). The products in $\sum_{\langle \text{inc.} \rangle}$ and $\sum_{\langle \text{decr.} \rangle}$ correspond to the remaining nearest neighbor bonds which are broken in the perpendicular and parallel needle orientation, respectively. The four terms in $\sum_{\langle \text{one} \rangle}^{(+)}$ correspond to the bonds between the external spin and those four lattice spins which are coupled to it in both the perpendicular and the parallel needle configuration. The sums $\sum_{\langle \text{decr.} \rangle}^{(+)}$ and $\sum_{\langle \text{inc.} \rangle}^{(+)}$ contain the terms which correspond to the rest of the bonds to the external spin in the perpendicular and the parallel needle orientation, respectively. As in the previous subsection the notation for these sums represents the modulus of their corresponding prefactors in the crossover Hamiltonian

$$\mathcal{H}_{\text{cr}}^{[+]}(\lambda) = \tilde{\mathcal{H}}^{[+]} - \lambda \sum_{\langle \text{inc.} \rangle} - (1-\lambda) \sum_{\langle \text{decr.} \rangle} - \sum_{\langle \text{one} \rangle}^{(+)} - \lambda \sum_{\langle \text{inc.} \rangle}^{(+)} - (1-\lambda) \sum_{\langle \text{decr.} \rangle}^{(+)} . \tag{4.17}$$

The first term $\tilde{\mathcal{H}}^{[+]} \equiv \mathcal{H}^s + \sum_{\langle \text{zero} \rangle} + \sum_{\langle \text{inc.} \rangle} + \sum_{\langle \text{decr.} \rangle}$ corresponds to a strip with a cross-shaped hole where the bonds belonging to $\sum_{\langle \text{zero} \rangle}$, $\sum_{\langle \text{inc.} \rangle}$, and $\sum_{\langle \text{decr.} \rangle}$ are absent. In Fig. 15(c) this means that the dashed and dash-dotted bonds and all the arrows are removed. In $\mathcal{H}_{\text{cr}}^{[+]}(\lambda)$ the contributions due to the two types of bonds missing in $\tilde{\mathcal{H}}^{[+]}$ (i.e., $\sum\limits_{\langle \text{inc.} \rangle}$ and $\sum\limits_{\langle \text{decr.} \rangle}$) contain prefactors $-\lambda$ and $-(1-\lambda)$ of increasing and decreasing strengths, respectively. Bond contributions which couple the lattice spins in $\sum\limits_{\langle \text{one} \rangle}^{(+)}$, $\sum\limits_{\langle \text{inc.} \rangle}^{(+)}$, and $\sum\limits_{\langle \text{decr.} \rangle}^{(+)}$ (with strengths 1, λ, and $1-\lambda$, respectively) are added to the external spin $s_0 = 1$. As λ increases from 0 to 1 Fig. 15(c) clearly illustrates the crossover from the perpendicular to the parallel needle orientation considered in Figs. 15(a) and 15(b).

4.4. Numerical results for the critical quasi-torque

In order to determine the dependence of the critical torque on v_N we have used system sizes 1000×100 and 1000×101 for $(b[O]t)$ and $(b[+]t)$ needles, respectively. We have used the hybrid Monte Carlo method for the sequential generation of system configurations. One MC step consists of Wolff cluster[59] update followed by $L \times W/4$ attempts of Metropolis flips[60] of randomly chosen spins and of additional $(D+3)^2$ flips of randomly chosen spins in the square of size $(D+3) \times (D+3)$ with the center at position $(0, v_N)$. We have used 1.5×10^7 MC steps for thermalization, followed by the computation of the thermal average using 8×10^7 MC steps. These latter MC steps have been split into 16 intervals to estimate the numerical inaccuracy. For every selected set of parameters (i.e., type of needle and boundary conditions $(b[n]t)$, the strip length L, the strip width W, the needle length D, and the separation of the needle from the midline v_N) we have performed computations for 32 values of the coupling parameter $\lambda_k = \frac{k}{31}$, $k = 0, 1, \ldots, 31$ and then we have carried out the numerical integration by using the extended version of Simpson's rule.

The *"small needle expansion"* (SNA) for non-spherical particle has been used for theoretic evaluation of the free energy difference. We use the expression (Eqs.(2.16)-(2.18) of Ref. [46])

$$\Delta F = \Delta F_l + \Delta F_{nl} + \ldots ,$$ (4.18)

where the leading contribution F_l does not depend on v_N

$$\frac{\Delta F_l}{k_B T} = -\pi \left(\frac{D}{2W} \right)^2 \Delta_{b,t}(W/L) .$$ (4.19)

Here, the values of the constant for BCs are $\Delta_{b,t} = (\pi/48)\{-1, -1, 23, 2\}$ for $(b,t)=\{(O,O), (+,+), (+,-), (+,O)\}$, respectively. In the cases $(b[O]t)$

the next-to-leading term reads

$$\frac{\Delta F_{nl}}{k_B T} = \frac{1}{256}\left(\frac{\pi D}{W}\right)^3 g_{b,t}, \tag{4.20}$$

which depends on v_N via the following simple expressions for $g_{b,t}$:

$$
\begin{aligned}
g_{O,O} &= 3(\cos V)^{-3} - (5/3)(\cos V)^{-1}, \\
g_{+,+} &= -g_{O,O}, \\
g_{+,O} &= -g_{O,+} = (\tan V)[3(\cos V)^{-2} + (1/3)], \\
g_{+,-} &= g_{-,+} = -[3(\cos V)^{-3} + (7/3)((\cos V)^{-1} - 4\cos V)],
\end{aligned}
\tag{4.21}
$$

with $V = \pi v_N/W$.

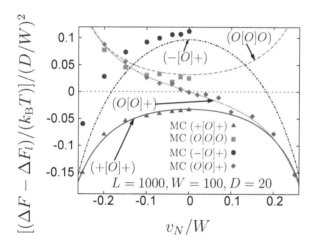

Figure 16. The normalized next-to-leading contribution $\Delta F_{nl} = \Delta F - F_l$ to quasi-torque ΔF acting on a small mesoscopic "ordinary" needle O of the length D as a function of normalized displacement v_N/W from the midline in a long strip of size $L \times W$ with corresponding boundaries $(b[O]t)$. MC results are plotted by symbols and "small needle" theoretical predictions are plotted by lines (from Ref. [46]).

In Fig. 16 we plot the data for $[(\Delta F - \Delta F_l)/(k_B T)]/(D/W)^2$ with ΔF obtained from a system with $L = 1000$, $W = 100$, and for the needle of the length $D = 20$. The coincidence with $[\Delta F_{nl}/(k_B T)]/(D/W)^2 \equiv c_{i,j}$ following from Eqs. (4.20) and (4.21) and shown by lines is very good. It means, ΔF is captured well by $\Delta F_l + \Delta F_{nl}$, except for the case $(-[O]+)$ in which it is still quite good. According to Eqs. (4.20) and (4.21) one has $c_{i,j} = (\pi^3/256)(D/W)g_{i,j}$. The lines are suitably normalized expressions c_{ij} (see Eq. (4.20)). There is very favorable agreement of the simulation data, which correspond to $W/L = 0.1$, with the analytic predictions for $W/L = 0$. Good agreement signals that all the mesoscopic distances and lengths, including D, chosen in the simulations turn out to be large enough for the lattice system to lie within the universal scaling region.

In Fig. 17 we compare simulation data for the cases of the normal needle [+] embedded in strips with at least one "normal" boundary $(-[+]O), (+[+]O), (+[+]-), (+[+]+)$, and $(-[+]-)$ with the corresponding analytic predictions "SNA" (Eqs.(2.16)-(2.18) of Ref. [46]) for various positions v_N. We plot the normalized quasi-torque ΔF as a function of normalized displacement from the midline v_N/W for various boundary conditions for the needle of the length $D = 21$ on the strip of the length $L = 1000$ and width $W = 101$. The statistical error of the simulation data is comparable with the symbol sizes. There is reasonable agreement of the

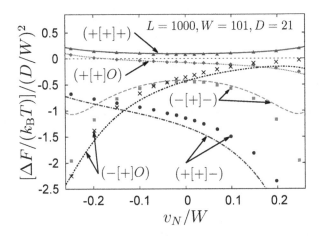

Figure 17. Normalized quasi-torque ΔF acting on a small mesoscopic "normal" needle [+] as a function of the normalized displacement v_N/W for various boundary conditions $(t[+]b)$.

simulation data for $W/L = 0.1$ with the analytic predictions of the "small needle approximation" (from Ref. [46]).

5. Numerical method for sphere-sphere geometry

Experiments on colloid aggregations[13–16] demonstrate appearance of attractive forces between particles in a critical media. It motivates us to apply a numerical method for computation of CC interactions for particle-particle

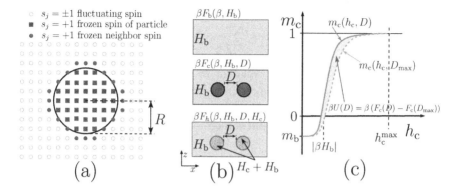

Figure 18.   (a) Schematic representation for a quasi-sphere on the lattice. (b) Computation of the insertion free energy: bulk system with the free energy $F_b(\beta)$, the system with fixed spins in two colloidal particles at distance D with $F_c(\beta, H_b, D)$, and the system with an external field H_c applied to spins of two colloidal particles at distance D with the free energy $F_h(\beta, H_b, D, H_c)$. (c) Typical graphs of magnetizations $m_c(H_c, D)$, $m_c(H_c, D_{max})$ as functions of 'colloid' field H_c for separations D, D_{max}. The shadowed area between curves is equal to the absolute value of the free energy difference $\beta U(D) = \beta F_c(D) - \beta F_c(D_{max})$.

geometry.   In this section we present results from Ref. [44]. As before, we consider the Ising model on a simple cubic lattice of size $L_x \times L_y \times L_z$ with periodic BCs. Our aim is to study the interaction between colloidal particles immersed in a critical binary mixture (Ising model universality class). Therefore we need the lattice representation of colloidal particles. Martin Hasenbusch has proposed[42] to draw a sphere of a certain radius R around a selected spin. All spins inside the sphere are considered to belong to the colloidal particle and fixed to be $+1$. In Fig. 18(a) we plot a cross section of a sphere of the radius $R = 3.5$, spins inside the sphere are denoted by filled squares. We consider the case of very strong positive surface fields for colloids. This choice corresponds to the symmetry-breaking BCs with completely ordered surface, i.e., $(+)$ (see Subsection 2.4 for details). It means, that a neighbor spin j, which is in a contact with a particle surface will be frozen $s_j = +1$, such spins are denoted by filled circles. We denote as {col} the set of all frozen spins in the system (spins in both colloidal particles and their neighbors, the total number is N_c spins) and refer to this set as spins of colloidal particles. In Fig. 18(a) we indicate these spins by filled symbols. Fluctuating spins in the bulk are represented by empty circles. We denote as F_b the free energy of an empty system with periodic BCs (see

Fig. 18(b) top) with the standard Hamiltonian for a spin configuration $\{s\}$

$$\mathcal{H}_{\mathrm{b}}(\{s\}) = -J\sum_{\langle ij \rangle} s_i s_j - H_{\mathrm{b}}\sum_n s_n, \tag{5.1}$$

where $J = 1$ is the interaction constant, H_{b} is the bulk magnetic field, the sum $\langle ij \rangle$ is taken over all neighbor spins, the sum over n runs over all spins of the spin configuration $\{s\}$. The free energy of the system is expressed via the sum over all possible spin configurations Ω as

$$F_{\mathrm{b}}(\beta, H_{\mathrm{b}}) = -\frac{1}{\beta}\log\left[\sum_{\{s\}\in\Omega} e^{-\beta\mathcal{H}_{\mathrm{b}}(\{s\})}\right]. \tag{5.2}$$

The Hamiltonian of the system with two colloidal particles of a radius R at a distance D (see Fig. 18(b) middle) is given by the same Eq. (5.1). But for this system all spins $s_k \in \{\mathrm{col}\}$ of colloidal particles and their neighbors $\{\mathrm{col}\}$ should be frozen $s_k = +1$, $k \in \{\mathrm{col}\}$, therefore the free energy is

$$F_{\mathrm{c}}(\beta, H_{\mathrm{b}}) = -\frac{1}{\beta}\log\left[\sum_{\{s\}\in\Omega} e^{-\beta\mathcal{H}_{\mathrm{b}}(\{s\})} \prod_{k\in\{\mathrm{col}\}} \delta_{s_k,1}\right]. \tag{5.3}$$

The product of the Dirac delta functions $\delta_{s_k,1}$ fixes the values of spins $s_k = +1$ in colloidal particles $k \in \{\mathrm{col}\}$. In this expression for a free energy we also count the interaction between frozen spins within particles. Let us introduce the system with the Hamiltonian

$$\mathcal{H}_{\mathrm{h}}(\{s\}) = -J\sum_{\langle ij \rangle} s_i s_j - H_{\mathrm{b}}\sum_n s_n - H_{\mathrm{c}}\sum_{k\in\{\mathrm{col}\}} s_k, \tag{5.4}$$

where the additional local magnetic field H_{c} is applied to spins s_k of colloidal particles $k \in \{\mathrm{col}\}$ as shown in Fig. 18(b), bottom. The free energy of this system is given by the formula

$$F_{\mathrm{h}}(\beta, H_{\mathrm{b}}, D, H_{\mathrm{c}}) = -\frac{1}{\beta}\log\left[\sum_{\{s\}\in\Omega} e^{-\beta\mathcal{H}_{\mathrm{h}}(\{s\})}\right]. \tag{5.5}$$

For zero additional field this free energy equals the free energy of a system without particles $F_{\mathrm{h}}(\beta, H_{\mathrm{b}}, D, H_{\mathrm{c}} = 0) = F_{\mathrm{b}}(\beta, H_{\mathrm{b}})$. Let us consider the system with certain bulk field H_{b} at fixed inverse temperature β. Therefore in the description of the numerical method arguments of functions (β, H_{b}) are omitted for the simplicity of notation. For a very strong value of the additional field $\beta H_{\mathrm{c}} \gg 1$ spins in colloidal particles $\{\mathrm{col}\}$ (the number of these spins is N_{c}) are frozen, therefore the free energy has a limit

$\lim\limits_{\beta H_c \to \infty} F_h(H_c, D) = F_c(D) - H_c N_c$. Let us introduce the local magnetization variable $h_c = \beta H_c$. One can express the magnetization of spins in
colloids $M_c = \sum\limits_{k \in \{\text{col}\}} s_k$ via the derivative of the free energy with respect
to h_c:

$$M_c(h_c, D) = -\frac{\partial[\beta F_h(h_c/\beta, D)]}{\partial h_c}. \tag{5.6}$$

We denote as $m_c(h_c, D) = M_c(h_c, D)/N_c$ the magnetization normalized per
total number N_c of spins in particles. The free energy is expressed as an
integral over the local magnetization variable

$$\beta F_h(H_c, D) = \beta F_b - N_c \int\limits_0^{\beta H_c} m_c(h_c, D)\mathrm{d}h_c. \tag{5.7}$$

For big enough maximal value of the additional field $h_c^{\max} \gg 1$ we can
express the free energy of the system with colloidal particles as

$$\beta F_c(D) = \beta F_b + N_c \int\limits_0^{h_c^{\max}} [1 - m_c(h_c, D)]\,\mathrm{d}h_c. \tag{5.8}$$

For this system the particles magnetization at zero additional field $h_c = 0$
equals the bulk magnetization $m_c(h_c = 0, D) = m_b$ and it is equal to
1 at strong $h_c \gg 1$ field $\lim\limits_{h_c \to \infty} m_c(h_c, D) = 1$. Therefore the result of the
integration in Eq. (5.8) does not depend on the upper limit of the integration
for big enough $h_c^{\max}$ (we use the value $h_c^{\max} = 5$). We schematically plot
the magnetization $m_c(h_c, D)$ for the case of the negative bulk magnetic
field $H_b < 0$ in Fig. 18(c). The "insertion" free energy $\beta F_c(D) - \beta F_b$
"graphically" equals to the area between lines $m_c(h, D)$ and 1.

We would like to compute the potential $\beta U(D)$ of the Casimir force
$f_C(D)$ (measured in $k_B T$ units) between two quasi-spherical particles at the
distance D. Up to a certain constant C_1 this potential may be expressed
via the free energy $\beta U(D) = \beta F_c(D) + C_1$. We select this constant equal
to the negative value of the free energy at some maximal separation $D_{\max}$:
$C_1 = -\beta F_c(D_{\max})$. The potential of CC interaction is provided by formula
$$\beta U(D) = N_c \int\limits_0^{h_c^{\max}} [m_c(h_c, D_{\max}) - m_c(h_c, D)]\mathrm{d}h_c.$$ Graphically, in Fig. 18(c),
this function equals to the area between lines $m_c(h_c, D)$ and $m_c(h_c, D_{\max})$
with the minus sign. This quantity is independent from details of spins
interaction inside particles because these contributions drop out from the

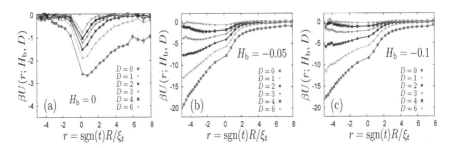

Figure 19. The Casimir interaction potential $U(r; H_b, D)$ as a function of the scaling variable $r = \text{sgn}(t)R/\xi_t$ for various values of the separation $D = 0, 1, 2, 3, 4, 6$ for: (a) zero bulk field $H_b = 0$, (b) negative bulk field $H_b = -0.05$, and (c) negative bulk field $H_b = -0.1$.

difference. The method described above is optimized for the computation of the potential of the Casimir interaction βU. For the computation of the Casimir force $f_C = -\frac{\partial[\beta U(D)]}{\partial D}$ between two particles we can compute the difference $f_C(D - 1/2) = N_c \int_0^{h_c^{\max}} [m_c(h_c, D) - m_c(h_c, D - 1)]dh_c$. It is also possible to modify the proposed method for the interpolation between two configurations for distances D and $D - 1$ by varying the local field H_c.

Numerical simulations have been performed for the system of the size $78 \times 49 \times 49$. Two quasi-spherical particles of radius $R = 3.5$ are located at separation D along the x direction as shown in Fig. 18(b) (the $x - z$ cross section). The particles are in the contact for separation $D = 0$. The distance $D_{\max} = 30$ is the maximal possible interparticle separation in the x direction for the system with periodic BCs. We use the histogram technique (see Appendix A.2) for the precise integration of the particle magnetization over the local magnetic field. The probability distribution $P(m_c, h_c)$ of the particle magnetization m_c is proportional to the exponent $\exp(h_c N_c m_c)$. We compute this probability distribution for 16 values of the additional field $h_c^j = \{0, 0.01, 0.02, 0.03, 0.04, 0.05, 0.07, 0.1, 0.16, 0.23, 0.4, 0.5, 0.7, 1, 1.5, 2.5\}$. The probability distribution for the value of the field h_c may be expressed as

$$P(m_c, h_c) = \frac{1}{A} \exp[(h_c - h_c^j)N_c m_c], \qquad (5.9)$$

where the normalization constant $A = \sum_{m_c} \exp[(h_c - h_c^j)N_c m_c]$. The values of fields should be close enough to let the probability distributions intersect.

We present results for two particular cases: the constant magnetic field and various temperatures and constant temperatures and various values of the magnetic field. For the constant magnetic field we use the scaling variable $r = \text{sgn}(t)R/\xi_t$. In Figs. 19(a)-19(c) we plot the interaction potential $\beta U(r; H_b, D)$ as a function of the scaling variable r for separations $D = 0, 1, 2, 3, 4, 6$ and values of the bulk field $H_b = 0, -0.05$, and -0.1, respectively. For zero bulk field Fig. 19(a) the attractive potential has a pronounced minimum in the vicinity of the critical point $r \simeq 0$. For the negative value of the bulk field $H_b = -0.05$ the amplitude of the attractive interaction increases several times. For small separations $D = 1, 2, 3 < R$ the minimum of the interaction disappears and the interaction within the investigated range $-4.5 < r < 8$ has no minima. For big enough separations $D = 4, 6 > R$ the width of the interaction potential well increases with respect to r. The strongest interaction corresponds to the smallest value of r. In Figs. 20(a)-20(c) we plot the CC interaction potential $\beta U(\eta; \beta, D)$

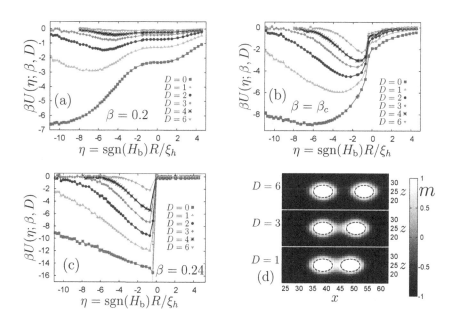

Figure 20. The Casimir interaction potential $\beta U(\eta; \beta, D)$ as a function of the bulk field scaling variable $\eta = \text{sgn}(H_b)R/\xi_h$ for various values of separation $D = 0, 1, 2, 3, 4, 6$: (a) above the critical point $\beta = 0.2$, (b) at the critical point $\beta_c \simeq 0.221654$, and (c) below the critical point $\beta = 0.24$. (d) The magnetization profile $m(x, z)$ as a function of coordinates x, z for $\beta = 0.25$ ($r \simeq -3.65$), $H_b = -0.1$ ($\eta \simeq -4.54$), and various separations $D = 1, 3, 6$.

as a function of the scaling variable $\eta = \text{sgn}(H_b)R/\xi_h$ for various separations and temperatures $\beta = 0.2, \beta_c$, and 0.24 (above T_c, at T_c, and below T_c, respectively). For $D > 0$ the interaction potential has a minimum as a function of η. The depth of this minimum decreases with increasing separation D. Above T_c the minima are smooth and shifted to stronger negative values of $\eta \sim -4, -6$. Below T_c the minima become sharp and narrow, shifted to smaller in the amplitude values of the negative field $h \sim -2$. In Fig. 20(d) we plot the magnetization profile $m(x, z)$ as a function of coordinates (x, z) for the value of the inverse temperature $\beta = 0.25$ (the corresponding value of the scaling variable $r \simeq -3.65$) and the value of the magnetic field $H_b = -0.1$ (the value of the scaling variable $\eta \simeq -4.54$ using the colormap. In Figs. 19(b),(c), and 20(a) one can see the "stretching" of graphs in the vicinity of zero due to the choice of the scaling variable r/ξ. We observe the formation of the bridge of positive spins (which corresponds to component A of the binary mixture) for small separations $D = 1, 3$. For larger separation $D = 6$ the bridge disappears which is consistent with the decreasing of the absolute value of the interaction potential. The strongest interaction for particles with $(+)$ boundary conditions (for colloidal particles with the surface that has a preference to A component) is observed for negative bulk fields $H_b < 0$ (B-rich phase of the binary mixture) below the critical point $T < T_c$ (above the lower critical point in the phase diagram Fig. 18(a)). For a small interparticle distances we observe the formation of the bridge of $+$ phase between particles that produces forces acting far away from criticality.

The proposed numerical method provides the potential energy difference for two interparticle distances D and $D_{\min}$. It is convenient for comparison with experimental results.[10,16] This potential has a simple graphical representation. The potential difference is proportional to the area between graphs of the local magnetization for these two separations. The proposed method may be also applied for studying of the multi-particle interactions in a critical solvent which plays significant role in the aggregation in the vicinity of the critical point.[79] Another advantage of this method is the nonzero bulk ordering field which mimics the shift of chemical potentials for components of the binary mixture.

6. Concluding remarks and perspectives

We give description of several methods for numerical simulation of Critical Casimir interactions. Generally speaking, all these approaches may be considered as variants of the common coupling parameter approach. The Hamiltonian of the system is represented as a sum of the basic part $\mathcal{H}_0$ plus the difference $\lambda \Delta \mathcal{H}$ that is proportional to the coupling constant λ. This coupling parameter crosses over the system from the initial to the final state. The derivative of the free energy $\beta F(\lambda) = -\ln\left[\sum e^{\beta(\mathcal{H}_0 + \lambda \Delta \mathcal{H})}\right]$ over the coupling parameter is the thermal average of the difference $\langle \Delta \mathcal{H} \rangle$. Integrating the latter over the λ we can compute the desirable free energy difference $F(1) - F(0)$.

For the "pure" approach the coupling parameter λ interpolates the Hamiltonian between two states of the system. This method has been used for the computation of the free energy difference in Refs. [24,26–29,43,46]. The bulk magnetic field or the local magnetic field also may play a role of the coupling parameter (see Sections 2, 5). In this case the total magnetization (for the case of the free energy density for a system with the bulk ordering field[29,41]) or local[44] magnetization (for the case of particle-particle interaction) represent the Hamiltonian difference $\Delta \mathcal{H}$. The initial and final states of the system are states with different values of the magnetic field. In the case of the energy integration method the inverse temperature itself plays a role of the coupling constant. The initial and final states of the system are states with different temperatures. This method has been used for the computation of the bulk free energy density[28,29] and for the computation of the free energy differences.[33–39]

We can also characterize described before methods by a number of spins N_m in measured observables. For example, we would like to measure the interaction force (or free energy difference) between two particles in a bulk. In this case the force weakly depends on the size of the system. For the temperature integration method the observable is the internal energy of the system and this quantity is proportional to the system size $N_m \propto L_x \times L_y \times L_z$. One can measure the difference between two quantities with fluctuations of order of $\sqrt{N_m}$. For the coupling parameter approach with removing a selected layer (see Section 3) $N_m \propto L_x \times L_y$ that decreases fluctuations of the observable. Finally, for the local field method (see Section 5) $N_m \propto N_c$, where N_c is the number of spins in particles and this quantity does not depend on the system size. Therefore from the point of view of efficiency, the latter method is preferable. One can easily adopt these methods for studying CC interactions between a patterned substrate and a particle, between particles of non-spherical shape etc..

A.1. Restoring the scaling function of the force difference

Using the coupling parameter approach we can compute the Casimir force difference $\Delta f_C(\beta, L_1, L_2) = f_C(\beta, L_1 - 1/2) - f_C(\beta, L_2 - 1/2)$ for a fixed system cross-section area $A = L_x \times L_y$ and two fixed values of the system thickness $L_z = L_1$, L_2. The scaling function ϑ of the Casimir force can be extracted from the temperature dependence of Δf_C by using the fact that f_C scales according to Eq. (1.1) for large $L_{1,2}$ and A. It is convenient to introduce the quantity

$$g(\tau; L_1, L_2) \equiv (L_1 - 1/2)^d \, \Delta f_C(\beta = \beta(\tau; L_1), L_1, L_2, A) \,, \qquad (\text{A.1})$$

as a function of $\tau = L^{1/\nu}(\beta_c - \beta)/\beta$, where $\beta(\tau; L) = \beta_c/[1 + \tau L^{-1/\nu}]$. According to Eq. (1.1), g is expected to scale as

$$g(\tau; L_1, L_2) = \vartheta(\tau) - \alpha^{-d}\vartheta(\alpha^{1/\nu}\tau) \,, \qquad (\text{A.2})$$

where $\alpha = (L_2 - 1/2)/(L_1 - 1/2)$ is the width ratio, and $\vartheta(\tau)$ is the Monte Carlo estimate of the scaling function ϑ, here $d = 3$ is the spatial dimension. Even though ϑ is independent of the geometrical realization of the simulation cell (in the limit $L \to \infty$, $A \to \infty$), in the general case the lattice estimate ϑ might depend on it via A and $L_{1,2}$ due to corrections to scaling. For a given pair of geometries $L_1 \times A$ and $L_2 \times A$, we can compute data for $\Delta f_C(\beta, L_1, L_2)$ at different temperatures and obtain $g(\tau; L_1, L_2)$ for a discrete set of values of τ. In order to determine $\vartheta(\tau)$ from the numerical data for $g(\tau; L_1, L_2)$ with fixed $L_{1,2}$ and A, one can solve Eq. (A.2) iteratively. One can expect that this provides a solution together with the property $\vartheta(|\tau| \to \infty) \to 0$ (or $\vartheta(\tau \to -\infty) \to const$, for the XY model below the critical point $T < T_c$). The method, described below, was used in Refs. [26,27] for the temperature scaling variable. Recently, this approach was applied for computation of the scaling function as a function of the bulk field scaling variable.[41]

As a first approximation of the actual $\vartheta(\tau)$ one may take $\vartheta_0(\tau) \equiv g(\tau; L_1, L_2)$, which can be improved by taking into account that Eq. (A.2) yields $\vartheta(\tau) = \vartheta_0(\tau) + \alpha^{-d}\vartheta(\alpha^{1/\nu}\tau) \simeq \vartheta_0(\tau) + \alpha^{-d}\vartheta_0(\alpha^{1/\nu}\tau)$. Accordingly, a better next step approximation $\vartheta_1(\tau)$ is provided by

$$\vartheta_1(\tau) = \vartheta_0(\tau) + \alpha^{-d}\vartheta_0(\alpha^{1/\nu}\tau). \qquad (\text{A.3})$$

The values of the function ϑ_0 at the point $\alpha^{1/\nu}\tau$ are obtained by the cubic spline interpolation of the data over available points. We can replace ϑ_0 in Eq. (A.3) by using Eq. (A.2), yielding $\vartheta_1(\tau) = \vartheta(\tau) - \alpha^{-2d}\vartheta(\alpha^{2/\nu}\tau) \simeq$

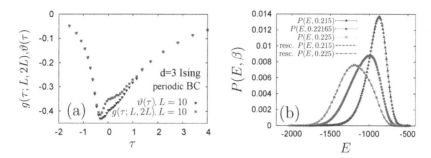

Figure A.1. (a) The force difference $g(\tau; L, 2L)$ (from Ref. [27]) and the associated restored scaling function $\vartheta(\tau)$ which is calculated by solving Eq. (A.1)) iteratively for the $d = 3$ Ising model with periodic BC and the base system thickness $L = 10$. (b) Probability distribution function $P(E, \beta)$ of the energy E for $\beta = 0.215, 0.22165, 0.2215$ computed via MC (triangles, circles, and squares) and rescaled for $\beta' = 0.215, 0.2215$ from the data for $\beta_i = 0.22165$ (dashed and dash-dotted lines) for the system of size $L_{\text{cube}} = 32$.

$\vartheta(\tau) - \alpha^{-2d}\vartheta_1(\alpha^{2/\nu}\tau)$, which indicates how the approximation $\vartheta_1(\tau)$ can be improved by introducing $\vartheta_2(\tau) = \vartheta_1(\tau) + \alpha^{-2d}\vartheta_1(\alpha^{2/\nu}\tau) = \vartheta(\tau) - \alpha^{-4d}\vartheta(\alpha^{4/\nu}\tau)$. This estimation can in turn be used to further improvement in the same way. The resulting iterative procedure describes a sequence of approximations which converges very rapidly

$$\vartheta_{k \geq 1}(\tau) = \vartheta_{k-1}(\tau) + \alpha^{-2^{k-1}d}\vartheta_{k-1}(\alpha^{2^{k-1}/\nu}\tau). \tag{A.4}$$

The correction to the k-th approximation is of the order of $\alpha^{-2^{k-1}d}$ therefore exponentially small in 2^k. In addition, $\vartheta(\tau)$ is generally expected to decay exponentially for large $|\tau|$. For $\alpha \simeq 2$, already for $k = 5$ one has the estimation $\alpha^{-2^{k-1}d} \simeq 3 \times 10^{-15}$ in three dimensions $(d = 3)$. The choice of the ratio $\alpha \simeq L_2/L_1$ is a compromise between two competing aims: a small value decreases the accuracy of a given approximation but on the other hand reduces the sizes of the lattices and, correspondingly, the computational time. Our choice of geometries $L_2 = 2L_1$ gives $\alpha \simeq 2$ and provides a very good approximation of $\vartheta(\tau) \equiv \vartheta_{k \to \infty}(\tau)$ already for $\vartheta_5(\tau)$.

To demonstrate the application of this restoring procedure, we plot in Fig A.1(a) g and ϑ as functions of τ for the Ising model with periodic BC. The basic system thickness $L = 10$, the length and width of the box are $L_x = L_y = 6L$. In this figure we do not take into account finite-size corrections to scaling which are of order of L^{-3} for periodic BC.[27] The pronounced shoulder in the vicinity of $\tau = 0$ originally present in g is smoothed out upon calculating the scaling function.

A.2. Histogram reweighting technique

One can apply the histogram reweighting technique[61,80] for determination of the bulk free energy. The idea of this method is to extract the value of the average energy $\langle E \rangle(\beta')$ at the inverse temperature β' from the probability distribution function $P(E, \beta_i)$ of the energy E at the inverse temperature β_i. Using the the Boltzmann statistical weight $P(E, \beta) \propto \exp(-\beta E)$ we obtain expression for the probability $P(E, \beta') = \frac{1}{C} P(E, \beta_i) e^{E(\beta_i - \beta')}$ where the constant $C = \sum_E P(E, \beta_i) e^{E(\beta_i - \beta')}$ provides the probability normalization $\sum_E P(E, \beta') = 1$. In Fig. A.1(b) we demonstrate the application of this method: rescaled histograms for $\beta' = 0.215, 0.225$ based on $\beta_i = 0.22165$ (dashed and dash-dotted lines) are in perfect agreement with results of standard MC simulations (triangles, squares) performed for $\beta = 0.215, 0.225$, respectively.

In order to perform the energy integration the computation of the energy distribution $P(E, \beta_i)$ has been performed for the choice of 256 points of $\beta_i \in [0, 0.3]$. One can obtain the estimate for $\langle E \rangle$ at the inverse temperature β' based on the histogram $P(E, \beta_i)$ for the inverse temperature β_i:

$$\langle E \rangle_{\beta_i}(\beta') = \frac{\sum_E E P(E, \beta_i) e^{-E(\beta' - \beta_i)}}{\sum_E P(E, \beta_i) e^{-E(\beta' - \beta_i)}}. \tag{A.1}$$

For every interval $\beta' \in [\beta_i, \beta_{i+1}]$ we define the interpolated internal energy

$$\langle E \rangle(\beta', L_{\text{cube}}) = \frac{\beta_{i+1} - \beta'}{\beta_{i+1} - \beta_i} \langle E \rangle_{\beta_i}(\beta') + \frac{\beta' - \beta_i}{\beta_{i+1} - \beta_i} \langle E \rangle_{\beta_{i+1}}(\beta'), \tag{A.2}$$

which then has been used for the energy integration.

A.3. Critical amplitude ξ_0^H

One can compute the value of the critical amplitude ξ_0^H using the scaling relations $Q_2 = \left(\frac{\xi_0^H}{\xi_0^+} \right)^{\frac{\gamma}{\nu}} \frac{C^+}{C^c}$, $C^c = \frac{B^c}{\delta}$, $R_\chi = \frac{C^+ B^{\delta-1}}{(B^c)^\delta}$, where the universal amplitude ratios are $Q_2 = 1.195(10)$ and $R_\chi = 1.662(5)$ (see Ref. [64] where ξ_0^H is denoted as f^c and ξ_0^+ is denoted as f^+). Substituting values $C^+ = 1.106(5)$[81] and $B = 1.6919045$[82] we obtain the critical amplitude $\xi_0^H = 0.278(2)$.

A.4. Fitting procedure for correction to scaling

In this subsection the general method that was used in order to obtain the best fitted values of the parameters which control the corrections to scaling is explained, see Refs. [26–29]. In the same way we have also computed the corrections due to a nonzero aspect ratio for the XY model. The standard way for the computation of the fitting parameters and for the evaluation of the associated confidence interval is the least square fit with chi-square tests of the quality. But in our case the additional complication is that the fitting function itself is not known and the information about this function should be extracted from the initial numerical data.

We would like to measure a quantity ψ, which in the limit of large system size L (i.e., in the absence of corrections to scaling) is expected to be a function of a scaling variable x only (which involves a suitable combination of the reduced temperature t and L of the system, e.g., $x = tL^{1/\nu}$ or $x = L/\xi$) so that $\psi(x, L \to \infty) = g(x)$, where g is the finite-size scaling function we would like to compute using MC simulations. We perform MC simulations for a set of N lattices of sizes L_1, L_2, ..., L_N and by varying the temperature we collect a discrete set of numerical values $\psi_{k,j}$ of ψ with $j = 1, 2 \ldots, j_k^{\max}$ which correspond to values $x_{k,j}$ of the scaling variable x in the interval $[x_{\min}, x_{\max}]$ for each size L_k. In this process the statistical inaccuracy $\Delta\psi_{k,j}$ associated with $\psi_{k,j}$ is also evaluated. From these values $\psi_{k,j}$ we should determine g. Due to the presence of corrections to scaling, ψ is actually not a function of a single variable x, but depends also of the system size L. In order to take corrections into account we assume the following functional structure (this form of the correction to scaling was used in Refs. [26,27]):

$$\psi(x; L) = f_1(L; a_1)g(f_2(L; a_2)x) \tag{A.1}$$

where $f_1(L; a_1)$ and $f_2(L; a_2)$ capture effects of the correction to scaling on the amplitude of ψ and on the scaling variable x. These functions depend on the system size L and on parameters a_1, a_2 which one would like to select achieving the best data collapse for the function g, obtained from the set of initial data points $(y_{k,j}(a_1, a_2), g_{k,j}(a_1, a_2)) = (f_2(L_k; a_2)x_{k,j}, \psi_{k,j}/f_1(L_k, a_1))$ for the various values of indexes j and k. The statistical inaccuracy for $g_{k,j}(a_1, a_2)$ is $\Delta g_{k,j}(a_1) = \Delta\psi_{k,j}/f_1(L_k, a_1)$.

For each value L_k in the interval $[x_{\min}, x_{\max}]$ we interpolated the data set $(x_{k,j}, \psi_{k,j})$ by a cubic spline approximation. This way we have constructed in the interval $[x_{\min}, x_{\max}]$ a continuous function $\psi_k(x)$ which passes through data points $\psi_k(x_{k,j}) = \psi_{k,j}$. From this function we have calculated the corresponding L_k-dependent evaluation $g_k(x; a_1, a_2)$ of g, given by formula

$$g_k(y; a_1, a_2) = f_1(L_k; a_1)^{-1}\psi_k(yf_2(L_k; a_2)^{-1}), \tag{A.2}$$

which fulfills $g_k(y_{k,j}(a_1, a_2); a_1, a_2) = g_{k,j}(a_1, a_2)$. In order to perform the fitting procedure we have to introduce the function g (which is still unknown) with which we would like to fit the data. For this purpose we define an expected function g_{expect} as the average over the various g_k,

$$g_{expect}(y; a_1, a_2) = \frac{1}{N} \sum_{k=1}^{N} g_k(y; a_1, a_2). \tag{A.3}$$

Then this function will be used for the χ^2 fitting procedure with initial MC data by adjusting parameters a_1 and a_2. Namely, we calculate the $\chi^2(a_1, a_2)$ associated with the fitting of the data points $(y_{k,j}(a_1, a_2), g_{k,j}(a_1, a_2))$ with the function $g_{expect}(y; a_1, a_2)$:

$$\chi^2(a_1, a_2) = \sum_{k=1}^{N} \sum_{j=1}^{j_k^{max}} \frac{[g_{k,j}(a_1, a_2) - g_{expect}(y_{k,j}(a_1, a_2); a_1, a_2)]^2}{[\Delta g_{k,j}(a_1)]^2}. \tag{A.4}$$

Due to the non-linear dependence of the fitted data and of the fitting function on the parameters a_i one cannot state that this quantity plays the same role as a χ^2 in more standard fitting procedures. Nevertheless, in order to determine the best fit parameters and the associated confidence intervals we have made an assumption that $\chi^2(a_1, a_2)$ is an analog of χ^2.

We determine the optimal fit parameters $\bar{a}_1$ and $\bar{a}_2$ which minimize the value of χ^2: $\chi^2(\bar{a}_1, \bar{a}_2) = \min_{\{a_1, a_2\}} \chi^2(a_1, a_2)$. The standard way for estimation of the statistical inaccuracy $\Delta \bar{a}_i$ for the fitting parameter $\bar{a}_i$ is to find the region of the plane (a_1, a_2) for which $\chi^2(a_1, a_2) < \chi^2(\bar{a}_1, \bar{a}_2) + 2.3$.[83] This region typically has the shape of an ellipse. Projections of this region onto the axis a_i gives $2\Delta \bar{a}_i$, so that the estimate for the parameters is of the form $\bar{a}_i \pm \Delta \bar{a}_i$. But usually, the result of this fitting procedure $\bar{a}_i$ depends on selection of the fitting interval $[x_{min}, x_{max}]$. Therefore, the final evaluation of the inaccuracy $\Delta \bar{a}_i$ for the fitting parameter a_i is typically larger, than the value extracted from the fitting procedure. The modification of this procedure for the case of a single fitting parameter a (e.g., L_{eff}) is straightforward.

A.5. Casimir force for Ginzburg-Landau model for planar geometry

In this subsection we formulate the computation of CCF within Ginzburg-Landau theory[19] (MFT universality class) which is exact for $d \geq 4$. Let us consider $d = 4$ dimensional slab of the thickness L and 3-dimensional cross-section A. The statistical weight of the order parameter field $\{\phi\}$ is $\exp(-\mathcal{H}(\{\phi\}))$, the Landau-Ginzburg Hamiltonian for planar geometry is

$$\begin{aligned}
\mathcal{H} = A \int_0^L \left[\frac{1}{2} \left(\frac{d\phi}{dz} \right)^2 + \frac{1}{2} t\phi^2 + \frac{g}{4!} \phi^4 - H_b \phi \right] dz + \\
+ A \left[\frac{1}{2} c_- \phi_0^2 - H_1^- \phi_0 + \frac{1}{2} c_+ \phi_1^2 - H_1^+ \phi_L \right],
\end{aligned} \tag{A.1}$$

where $t = (T - T_c)/T_c$, the constant $g > 0$, H_b is the bulk field. Surface fields H_1^- and H_1^+ are applied at bottom $-$ and top $+$ boundaries $\phi_0 = \phi(z = 0)$ and $\phi_L = \phi(z = L)$, and $1/c_-$, $1/c_+$ are extrapolation lengths. Within MFT, the temperature dependent ($H_b = 0$) correlation length $\xi_t = t^{-1/2}$ for $t > 0$, and $\xi_t = \frac{1}{\sqrt{2}}(-t)^{-1/2}$ for $t < 0$. The solution of the appropriate Euler-Lagrange equation

$$\frac{d^2\phi}{dz^2} - t\phi(z) - \frac{g}{6}\phi^3(z) + H_b = 0 \tag{A.2}$$

provides the equilibrium order parameter profile with the BC

$$\left. \frac{d\phi}{d\tilde{z}} \right|_{\tilde{z}=0} = c_{\pm} \phi(\tilde{z} = 0) - H_1^{\pm}, \tag{A.3}$$

where $\tilde{z}$ is the separation from the wall, *i.e.*, $\tilde{z} = z$ for $-$ and $\tilde{z} = L - z$ for $+$. For large $H_1^{\pm}$, the leading behavior of $\phi(\tilde{z} \ll L)$ is given by $\sqrt{12/g}\tilde{z}^{-1}$ for $(+)$ BC and by $-\sqrt{12/g}\tilde{z}^{-1}$ for $(-)$ BC.[84] For large $c_{\pm}$ one has $\phi(\tilde{z} = 0) = 0$ corresponding to (O) BC. The CCF per unit area in units of $k_B T$ equals

$$f_C(t, H_b, L) = T_{zz}(z_0, t, H_b) - T_{zz}^b(t, H_b). \tag{A.4}$$

The stress tensor is

$$T_{zz}(z, t, H_b) = \frac{1}{2} \left(\frac{d\phi}{dz} \right)^2 - \frac{1}{2} t\phi^2(z) - \frac{g}{4!} \phi^4(z) + H_b \phi(z), \tag{A.5}$$

where z_0 is an arbitrary point $0 \leq z_0 \leq L$. The bulk contribution is

$$T_{zz}^b(t, H_b) = -\frac{1}{2} t\phi_b^2 - \frac{g}{4!}\phi_b^4 + H_b\phi_b \tag{A.6}$$

with ϕ_b as the solution of $t\phi_b + \frac{g}{6}\phi_b^3 = H_b$. For comparison, in Figs. 6(a), 6(b) at zero bulk field $H_b = 0$ the MFT results are plotted as functions of the scaling variable $x = \text{sgn}(t)(L/\xi_t)^\nu = tL$ and in Figs. 8(a)-8(d) at the critical point $t = 0$ as functions of the scaling variable $\text{sgn}(H_b)L/\xi_H$, where the bulk field depended correlation length is $\xi_H = \frac{1}{\sqrt{3}}|H_b|^{-1/3}$. The Casimir force scaling functions for Ginzburg-Landau theory and $(++)$ and $(-+)$ BC have been studied in Ref. [69].

References

1. H. B. Casimir, Proc. K. Ned. Akad. Wet. **51**, 793 (1948).
2. M. E. Fisher and P. G. de Gennes, C. R. Acad. Sci. Paris Ser. B **287**, 207 (1978).
3. M. Krech, *Casimir Effect in Critical Systems* (World Scientific, Singapore, 1994).
4. A. Gambassi, Journ. Phys.: Conf. Ser. **161**, 012037 (2009).
5. D. Beysens and D. Estève, Phys. Rev. Lett. **54**, 2123 (1985).
6. D. Beysens and T. Narayanan, Journ. Stat. Phys. **95**, 997 (1999).
7. R. Garcia and M.H.W. Chan, Phys. Rev. Lett. **83**, 1187 (1999).
8. A. Ganshin, S. Scheidemantel, R. Garcia, and M. H. W. Chan, Phys. Rev. Lett. **97**, 075301 (2006).
9. M. Fukuto, Y. F. Yano and P. S. Pershan, Phys. Rev. Lett. **94**, 135702 (2005).
10. C. Hertlein, L. Helden, A. Gambassi, S. Dietrich, and C. Bechinger, Nature **451**, 172 (2008).
11. A. Gambassi, A. Maciołek, C. Hertlein, U. Nellen, L. Helden, C. Bechinger, and S. Dietrich, Phys. Rev. E **80**, 061143 (2009).
12. U. Nellen, L. Helden, and C. Bechinger, Europhys. Lett. **88**, 26001 (2009).
13. S. Buzzaccaro, Journ. Colombo, A. Parola, and R. Piazza, Phys. Rev. Lett. **105**, 198301 (2010).
14. R. Piazza, S. Buzzaccaro, A. Parola, and J. Colombo, Journ. Phys.: Condens. Matter **23**, 194114 (2011).
15. S.J. Veen, O. Antoniuk, B. Weber, M.A.C. Potenza, S. Mazzoni, P. Schall, and G.H. Wegdam, Phys. Rev. Lett. **109**, 248302 (2012).
16. V.D. Nguyen, S. Faber, Z. Hu, G.H. Wegdam, and P. Schall, Nature Comm. **4**, 1584 (2013).
17. M. N. Barber, in *Phase Transitions and Critical Phenomena*, edited by C. Domb and J. L. Lebowitz (Academic, New York, 1983), Vol. 8.
18. V. Privman, in *Finite Size Scaling and Numerical Simulation of Statistical Systems*, edited by V. Privman (World Scientific, Singapore, 1990).
19. H. W. Diehl, in *Phase Transitions and Critical Phenomena*, edited by C. Domb and J. L. Lebowitz (Academic, London, 1986), Vol. 10, p. 76.
20. N.B. Wilding, Phys. Rev. E **55**, 6624 (1997).
21. S.L. Veatch, P. Cicuta, P. Sengupta, A. Honerkamp-Smith, D. Holowka, and B. Baird, ACS Chem. Biol. **3**, 287 (2008).
22. L. Onsager, Phys. Rev. **65**, 117 (1944).
23. B. Kaufman, Phys. Rev. **76**, 1232 (1949).
24. M. Krech and D.P. Landau, Phys. Rev. E **53**, 4414 (1996).
25. D. Dantchev and M. Krech, Phys. Rev. E **69**, 046119 (2004).
26. O. Vasilyev, A. Gambassi, A. Maciołek, and S. Dietrich, Europhys. Lett. **80**, 60009 (2007).
27. O. Vasilyev, A. Gambassi, A. Maciołek, and S. Dietrich, Phys. Rev. E **79**, 041142 (2009).
28. O. Vasilyev, A. Maciołek, and S. Dietrich Phys. Rev. E **84**, 041605 (2011).
29. O. Vasilyev and S. Dietrich, Europhys. Lett. **104**, 60002 (2013).

30. F.P. Toldin and S. Dietrich, Journ. Stat. Mech. P11003 (2010).
31. F.P. Toldin, M. Tröndle, and S. Dietrich, Phys. Rev. E **88**, 052110 (2013).
32. F.P. Toldin, M. Tröndle, and S. Dietrich, preprint arXiv:1409.5536 (2014).
33. A. Hucht, Phys. Rev. Lett. **99**, 185301 (2007).
34. A. Hucht, D. Grüneberg, and F.M. Schmidt, Phys. Rev. E **83**, 051101 (2011).
35. M. Hasenbusch, Journ. Stat. Mech. P07031 (2009).
36. M. Hasenbusch, Phys. Rev. B **81**, 165412 (2010).
37. M. Hasenbusch, Phys. Rev. B **82**, 104425 (2010).
38. M. Hasenbusch, Phys. Rev. B **85**, 174421 (2012).
39. M. Hasenbusch, Phys. Rev. B **83**, 134425 (2011).
40. M. Hasenbusch, Phys. Rev. E **80**, 061120 (2009).
41. D.L. Cardozo, H. Jacquin, and P.C.W. Holdsworth, Phys. Rev. B **90**, 184413 (2014).
42. M. Hasenbusch, Phys. Rev. E **87**, 022130 (2013).
43. O. Vasilyev and A. Maciołek, Journal of Non-Crystalline Solids **407**, 376 (2015).
44. O.A. Vasilyev, Phys. Rev. E **90**, 012138 (2014).
45. B.B. Machta, S.L. Veatch, and J.P. Sethna, Phys. Rev. Lett. **109**, 138101 (2012).
46. O. A. Vasilyev, E. Eisenriegler, and S. Dietrich, Phys. Rev. E **88**, 012137 (2013).
47. H. Hobrecht and A. Hucht, Europhys. Lett. **106**, 56005 (2014).
48. J.R. Edison, N. Tasios, S. Belli, R. Evans, R. van Roij, and M. Dijkstra, preprint arXiv:1403.4872 (2014).
49. M. Tröndle, S. Kondrat, A. Gambassi, L. Harnau and S. Dietrich, Europhys. Lett. **88**, 40004 (2009).
50. M. Tröndle, S. Kondrat, A. Gambassi, L. Harnau and S. Dietrich, Journ. Chem. Phys. **133**, 074702 (2010).
51. S. Kondrat, L. Harnau, and S. Dietrich, Journ. Chem. Phys. **131**, 204902 (2009).
52. T.F. Mohry, S. Kondrat, A. Maciołek, and S. Dietrich, Soft Matter **10**, 5510 (2014).
53. T.G. Mattos, L. Harnau, and S. Dietrich, Journ. Chem. Phys. **138**, 074704 (2013).
54. T.G. Mattos, L. Harnau, and S. Dietrich, preprint arXiv:1408.7081 (2014).
55. K. Binder, in *Phase Transitions and Critical Phenomena*, edited by C. Domb and J. L. Lebowitz (Academic, London, 1983), Vol. 8, p. 1.
56. H. W. Diehl, Int. Journ. Mod. Phys. B **11**, 3503 (1997).
57. K. K. Mon, Phys. Rev. B **39**, 467 (1989).
58. K. K. Mon and K. Binder, Phys. Rev. B **42**, 675 (1990).
59. U. Wolff, Phys. Rev. Lett. **62**, 361 (1989).
60. N. Metropolis and S. Ulam, Journ. Amer. Stat. Assoc. **44**, 335 (1949).
61. D. P. Landau and K. Binder, *A Guide to Monte Carlo Simulations in Statistical Physics* (Cambridge University Press, London, 2005), p. 155.
62. C. Ruge, P. Zhu, and F. Wagner, Physica A **209**, 431 (1994).
63. M. Hasenbusch, Phys. Rev. B **82**, 174433 (2010).

64. A. Pelissetto and E. Vicari, Phys. Rep. **368**, 549 (2002).
65. R. Guida and J. Zinn Justin, Journ. Phys. A: Math. Gen. **31**, 8103 (1998).
66. J. Engels, L. Fromme, and M. Seniuch, Nucl. Phys. B **655**, 277 (2003).
67. A. P. Gottlob and M. Hasenbusch, Physica A **201**, 593 (1993).
68. D. Grüneberg and H. W. Diehl, Phys. Rev. B **77**, 115409 (2008).
69. M. Krech, Phys. Rev. E **56**, 1642 (1997).
70. R. Evans and J. Stecki, Phys. Rev. B **49**, 8842 (1994).
71. Z. Borjan and P. J. Upton, Phys. Rev. Lett. **101**, 125702 (2008).
72. A. Maciołek, A. Drzewiński, and A. Ciach, Phys. Rev. E **64**, 026123 (2001).
73. M. Zubaszewska, A. Maciołek, and A. Drzewiński, Phys. Rev. E **88**, 052129 (2013).
74. B.V. Derjaguin, Kolloid Zeitschrift **69**, 155 (1934).
75. A. Hanke, F. Schlesener, E. Eisenriegler, and S. Dietrich, Phys. Rev. Lett. **81**, 1885 (1998)
76. D. Lingwood and K. Simons, Science **327**, 46 (2010).
77. S.L. Veatch and S.L. Keller, Biochim. Biophys. Acta, Mol. Cell Res. **1746**, 172 (2005).
78. B.B. Machta, S. Papanikolaou, J.P. Sethna, and S.L. Veatch, Biophys. Journ. **100**, 1668 (2011).
79. M.T. Dang, A.V. Verde, V.D. Nguyen, P.G. Bolhuis, and P. Schall, Journ. Chem. Phys. **139**, 094903 (2013).
80. Alan M. Ferrenberg and Robert H. Swendsen, Phys. Rev. Lett. **63**, 1195 (1989).
81. I.A. Campbell and P.H. Lundow, Phys. Rev. B **83**, 014411 (2011).
82. M. Caselle and M. Hasenbusch, Journ. Phys. A: Math. Gen. **30**, 4963 (1997).
83. P. R. Bevington, *Data Reduction and Error Analysis for the Physical Sciences* (McGraw-Hill, New York, 1969).
84. F. Schlesener, A. Hanke, and S. Dietrich, Journ. Stat. Phys. **110**, 981 (2003).

Chapter 3

Non-Ergodicity and Ageing in Anomalous Diffusion

Ralf Metzler*

*Institute of Physics & Astronomy, University of Potsdam,
Karl-Liebknecht Str. 24/25, 14476 Potsdam-Golm, Germany*

*Department of Physics, Tampere University of Technology,
Korkeakoulunkatu 10, FI-33720 Tampere, Finland*

Anomalous diffusion, the deviation from classical Brownian motion with its linear-in-time mean squared displacement, is a widespread phenomenon in a large variety of complex systems, ranging from amorphous semiconductors to biological cells. Anomalous diffusion processes are no longer confined by the central limit theorem, that enforces the convergence of Brownian motion to the Gaussian shape of the probability density function. Instead, anomalous processes have a rich variety of physical origins and non-traditional mathematical properties. We here provide a summary of various anomalous diffusion models, paying special attention to their non-ergodic and ageing behaviour. Whether a process is ergodic or not is of vital importance when measurements are evaluated in terms of time averaged observables such as the mean squared displacement. For non-ergodic processes these can no longer be interpreted by comparison to the typically calculated ensemble averaged observables. Breaking of ergodicity occurs in many anomalous diffusion processes. We also consider their ageing properties, the explicit dependence of observables on the time span between the original system initiation and the start of the measurement.

Contents

---

*rmetzler@uni-potsdam.de

1. Brownian motion

In 1828 Robert Brown—the Scottish botanist—reported observations of small granules of $\frac{1}{4000}$th to $\frac{1}{5000}$th of an inch extracted from larger pollen grains. He found these particles *evidently in motion*.[1] In a string of control experiments—even involving a bruised fragment of the Sphinx—Brown made meticulously sure that the motion he observed was not the effect of living matter, of *animalcules*. There exist reports of diffusive, erratic, jiggling motion of small particles before Brown's, notably those of Roman poet-philosopher Titus Lucretius in his book *De rerum natura* (on the nature of things) dated 50 BCE,[2] and Dutch physician Jan Ingenhousz in 1785.[3] However, Brown's name became epitomised as the synonym for thermal diffusion.

Adolf Fick in 1855 derived the diffusion equation (Fick's second law) from a combination of the continuity equation and the constitutive law (Fick's first law), in which the diffusive particle flux relates to the gradient of the particle concentration.[4] Albert Einstein's groundbreaking contribution to the theory of diffusion was its interpretation as a molecular rather than a continuum flux in a mean free path approach published in 1905.[5] This allows the interpretation of diffusion as a single particle process and the formulation of the diffusion equation for the probability density function (PDF) $P(\mathbf{r}, t)$ in the form

$$\frac{\partial}{\partial t} P(\mathbf{r}, t) = K_1 \nabla^2 P(\mathbf{r}, t), \qquad (1.1)$$

in which K_1 is the diffusion coefficient of physical dimension $[K_1] =$

cm^2/sec. ∇ is the gradient operator in d spatial dimensions. Einstein's second major contribution was the derivation of the relation

$$K_1 = \frac{k_B T}{m\eta} \tag{1.2}$$

between the diffusion coefficient and thermal energy $k_B T$, the mass m of the diffusing particle and the solution viscosity η. In Marian Smoluchowski's 1906 work the connection between the diffusion process and the mathematical approach using the random walk concept (originally formulated by Karl Pearson in 1905[6]) was then made more precise.[7] Paul Langevin [8] in 1908 then added the interpretation of the diffusion process in terms of a Newton equation with an additional, fluctuating force representing the erratic bombardment of a test particle by other particles in its environment. Relation (1.2) was also found by Australian physicist William Sutherland in the same year 1905.[9]

Solving the diffusion equation (1.1) for an initial particle position at the origin $\mathbf{r} = 0$, the Gaussian

$$P(\mathbf{r}, t) = \frac{1}{(4\pi K_1 t)^{d/2}} \exp\left(-\frac{\mathbf{r}^2}{4K_1 t}\right) \tag{1.3}$$

is obtained. Mathematically, this is a consequence of the central limit theorem. Namely, the random variable x, corresponding to a sum of increments of the underlying random motion—directed erratically to left and right— converges *a forteriori* to the normal, Gaussian distribution, as already anticipated by Thorvald Thiele in his work on the least squares method.[10] The mean squared displacement (in the ensemble sense) follows from the Gaussian (1.3) by integration, producing the famed linear time dependence

$$\langle \mathbf{r}^2(t) \rangle = \int_{-\infty}^{\infty} \mathbf{r}^2 P(\mathbf{r}, t) d\mathbf{r} = 2dK_1 t. \tag{1.4}$$

Inspired by the findings of Einstein, Smoluchowski, and also Stokes, Jean Perrin used the relation (1.2) between the microscopic quantity of the diffusion coefficient and the macroscopic value of thermal energy ($k_B T$ is equal to the value RT/N_A involving the gas constant R and Avogadro's number N_A) as a novel method to determine Avogadro's number. In 1908 Perrin reported the first systematic single particle tracking results in his seminal paper on diffusion. Perrin in his measurements traced small putty particles.[11] Due to the relatively short trajectories, Perrin fitted the distance distribution of many particles in a given time interval by the Gaussian PDF. This involved averaging over not completely identical particles,[11] a

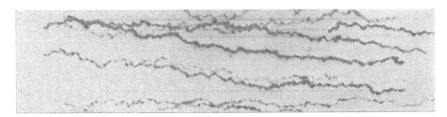

Figure 1. Sample trajectories of sedimenting mercury droplets measured by Ivar Nordlund in 1914 with his moving film technique. Here time increases to the right.[12] The
jiggly motion of the droplets superimposed onto the deterministic, linear sedimentation
shows the Brownian motion of the droplets.

drawback resolved only six years after Perrin's first publication and exactly
hundred years ago, in 1914 by the Swedish scientist Ivar Nordlund. Nordlund conceived an experimental setup using a stroboscope and a moving
film plate to record long time traces of small mercury droplets. From the
single droplet traces he then evaluated *time averages* of the mean squared
displacement.[12] Fig. 1 features some typical trajectories measured by Nordlund.[12] This method was continuously refined to culminate in the high precision measurements of Eugen Kappler.[13] Kappler's result for Avogadro's
number is within 1% of the best known value.

Today single particle tracking is a routine tool in complex fluids *in
vitro* as well as in living biological cells.[14] Common tracer particles include
fluorescently labelled molecules such as messenger RNA in the cytosol of
biological cells[15,16] or protein channels in the cell plasma membranes.[17]
Without labels, submicron tracers such as endogenous lipid[18] or insulin[19]
granules or internalised particles such as viruses[20] or plastic[21,22] spheres
can be directly monitored in microscopes. How can we evaluate the time
average of such single particle tracking experiments in terms of the theory
of diffusion involving the ensemble average (1.4)?

Single particle tracking experiments produce the time series $\mathbf{r}(t)$ of the
particle position. Typically, few but long trajectories $\mathbf{r}(t)$ are measured and
analysed based on the time averaged mean squared displacement (MSD)

$$\overline{\delta^2(\Delta)} = \frac{1}{T-\Delta} \int_0^{T-\Delta} \left[\mathbf{r}(t+\Delta) - \mathbf{r}(t)\right]^2 dt. \qquad (1.5)$$

This moving average integrates (in a discretised version, sums) the particle
displacements separated by the lag time Δ over the time series $\mathbf{r}(t)$ of length
(measurement time) T. For normal Brownian motion, the long time limit

produces[23] for

$$\overline{\delta^2(\Delta)} \simeq K_1 \Delta. \tag{1.6}$$

We thus find the equivalence

$$\langle \mathbf{r}^2(\Delta) \rangle = \overline{\delta^2(\Delta)} \tag{1.7}$$

of ensemble and long time averaged MSDs. More formally, we should write $\langle \mathbf{r}^2(\Delta) \rangle = \lim_{T \to \infty} \overline{\delta^2(\Delta)}$. This is in fact a restatement of Boltzmann's ergodic hypothesis known from statistical mechanics: long time averages and ensemble averages of physical observables are equivalent. In the following we will also consider the average over individual trajectories,

$$\left\langle \overline{\delta^2(\Delta)} \right\rangle = \frac{1}{N} \sum_{i=1}^{N} \overline{\delta_i^2(\Delta)} = \frac{1}{T - \Delta} \int_0^{T-\Delta} \left\langle \left[\mathbf{r}(t + \Delta) - \mathbf{r}(t) \right]^2 \right\rangle dt. \tag{1.8}$$

For finite particle traces, this trajectory-to-trajectory average smoothens the scatter present in individual traces.

2. Anomalous diffusion

Already in 1926 an exception to the linear time dependence (1.4) of the MSD of Brownian motion was analysed by English mathematician Lewis Fry Richardson. For the relative motion of two tracer particles in a driven turbulent flow he observed strongly non-Brownian behaviour.[24] Noticing that a particle cloud in this case does not spread like a Gaussian but fragments into a disjoint hierarchy of clusters, he introduced the notion of *non-Fickian diffusion* and used a diffusion equation with separation dependent diffusivity,

$$\frac{\partial q}{\partial t} = \text{const} \times \frac{\partial}{\partial l} \left[l^{4/3} \frac{\partial q}{\partial l} \right] \tag{2.1}$$

for the PDF $q(l, t)$ of the relative particle displacement. We will later come back to generalised diffusion equations of this type. For the MSD of the particle-particle separation l Richardson then obtained the power-law MSD $\langle l^2(t) \rangle \simeq t^3$ with the characteristic cubic scaling. Today *anomalous diffusion*[28] typically refers mostly to the generic power-law form[a]

$$\langle x^2(t) \rangle \simeq K_\alpha t^\alpha \tag{2.2}$$

of the MSD with the anomalous diffusion exponent α and the generalised diffusion coefficient K_α of physical dimension cm^2/sec^α. This is what we refer to in the following, distinguishing subdiffusion ($0 < \alpha < 1$) and superdiffusion ($\alpha > 1$). Anomalous diffusion is often measured in crowded media, in particular, in living biological cells.[29–33]

The conditions assumed by Einstein in his derivation of the diffusion equation are (i) the independence of individual particles, (ii) the existence of a sufficiently small time scale beyond which individual displacements are statistically independent, and (iii) the property that the particle displacements during this time scale correspond to a typical mean free path distributed symmetrically in positive or negative directions. These assumptions, by help of the central limit theorem, a forteriori lead to the Gaussian PDF (1.3) and thus to the diffusion equation (1.1). The model described by Einstein may therefore be viewed as a *random walk* or drunkard's walk, a concept introduced in the same year 1905 by Karl Pearson in his famed letter to Nature.[6] The connection of the diffusion law to the random walk process was rendered more precisely by Smoluchowski.[7]

In anomalous diffusion processes, at least one of these fundamental assumptions is violated, and the strong convergence to the Gaussian according to the central limit theorem broken. In particular, by departing from one or more of the assumptions (i)–(iii), we find that there exist a number of non-trivial generalisations of the Einstein-Smoluchowski diffusion picture, that is, anomalous diffusion loses the universality of Brownian motion, and the MSD (2.2) is no longer sufficient to uniquely identify a stochastic process. Many different stochastic processes give rise to anomalous diffusion, and they exhibit many different features. The question we address here is the violation of ergodicity: which anomalous diffusion processes encode ergodic behaviour (1.7), such that time averaged quantities—e.g., the time averaged MSD—can be evaluated in terms of their ensemble averaged analogues, which are typically available in physical theories? And in which cases is this ergodicity violated, and we need to provide new results for

---

[a]Curiously the very notion *anomalous diffusion* first appears in literature in the same year as Richardson's paper, 1926, but in the context of α rays.[25] Subsequently anomalous diffusion was used to describe the observation that in certain systems the 'Oeholm method' does not return a constant for the diffusivity as expected if the system were following Fick's law.[26] This paper by Herbert Freundlich and Deodata Krüger[26] refer to first measurements *in aqueous solutions of dyestuffs* by Herzog and Polotzky.[27]

these time averages such that the measured time series can be properly evaluated? As we will see, most anomalous diffusion processes in fact give rise to the disparity

$$\langle \mathbf{r}^2(\Delta) \rangle \neq \lim_{T \to \infty} \overline{\delta^2(\Delta)}, \qquad (2.3)$$

and accompanying properties. Jean-Philippe Bouchaud called stochastic processes featuring the inequivalence (2.3) weakly non-ergodic, to distinguish them from cases, in which the phase space is strictly disjoint, and a particle released in one part of the phase space can never reach another part (strongly non-ergodic).[34] In the cases discussed here it is the nonstationary character of the processes that gives rise to the disparity (2.3), while an ensemble of particles still evenly fills the phase space for sufficiently long trajectories.

We will start our journey through different popular anomalous diffusion models with the ergodic fractional Brownian motion and fractional Langevin equation motion in section 3. The subsequent sections then deal with weakly non-ergodic anomalous diffusion processes, starting with the classical example of subdiffusive continuous time random walks (CTRWs) with diverging characteristic waiting time in section 4. The following section 5 then deals with the experimentally relevant case when the anomalous motion of the particle includes an additional noisy component. Section 6 then considers the case of an ageing environment giving rise to ultraslow motion. The renewal character of continuous time random walk processes is shed off in section 7, in which we consider a simple model for correlations between successive waiting times or jump lengths. Superdiffusive continuous time random walks (Lévy flights and walks) are then addressed in section 8. Weak ergodicity breaking is then revealed in seemingly simple processes, namely, the two approaches used to explain the above cubic Richardson diffusion. Thus, section 9 considers scaled Brownian motion with an explicitly time dependent diffusivity (corresponding to Batchelor's approach), while section 10 uses a position dependent diffusion coefficient (generalising Richardson's model). Ultraslow diffusion in quenched environments (Sinai diffusion) and ultraslow continuous time random walk processes are addressed in section 11. Many-body interactions combined with renewal, power-law waiting times are shown to give rise to ultraslow motion of a labelled particle in section 12. Finally, we mention the diffusion on a fractal medium in section 13, before concluding in section 14.

3. Fractional Brownian and Langevin equation motion

The well known Langevin equation in the overdamped limit[b]

$$\frac{dx(t)}{dt} = \sqrt{2K_1} \times \xi(t) \tag{3.1}$$

is driven by white Gaussian noise of zero mean and correlator $\langle \xi(t)\xi(t')\rangle \sim \delta(t - t')$.[8,35] In contrast to the δ-correlation fractional Gaussian noise (fGn) Gaussian has the power-law correlation

$$\langle \xi(t)\xi(t')\rangle \sim \alpha K_\alpha(\alpha - 1)|t - t'|^{\alpha-2}, \tag{3.2}$$

with exponent $0 < \alpha < 2$. FGn is known to characterise the tracer motion in viscoelastic environments.[16,18,19,36–38] Such correlated noise also governs the motion of individual lipids in lipid membranes,[39,40] and fGn occurs for the motion of a tracer particle in a single file of colloidal particles with excluded volume interactions .[41,42] In the case $0 < \alpha < 1$ the noise-noise correlator has a negative sign, a situation often termed antipersistent noise. In the case $1 < \alpha < 2$ we speak of persistence.

3.1. Fractional Brownian motion (FBM)

Fractional Brownian motion simply substitutes the white Gaussian noise in the Langevin equation (3.1) with fGn (3.2).[43,44] From a physical point of view, fGn is to be considered an external noise. The resulting ensemble average for the MSD is given by Eq. (2.2). FBM is ergodic in the sense that the time averaged MSD for unconfined motion becomes[45]

$$\overline{\delta^2(\Delta)} \sim 2K_\alpha \Delta^\alpha \tag{3.3}$$

in the limit of long T. We emphasise that the equality $\overline{\delta^2(\Delta)} = \langle x^2(\Delta)\rangle$ indeed holds for a single trajectory in the long measurement time T limit,[36] as expected for an ergodic process. The approach to ergodicity occurs as a power-law, similar to regular Brownian motion.[45]

In addition to the ergodic behaviour, individual trajectories of FBM are reproducible. More precisely, the amplitude variation of the time averaged MSD $\overline{\delta^2(\Delta)}$ from different realisations of length T around the mean $\langle \overline{\delta^2(\Delta)}\rangle$ is Gaussian. At a fixed lag time Δ, the width of this distribution decreases with increasing measurement time T,[46] and sufficiently long individual trajectories are therefore in that sense reproducible.

---

[b]For simplicity, we will use the one-dimensional notation for the remainder of this chapter.

3.2. *Fractional Langevin equation motion*

When we require that the fGn is internal and should fulfil the Kubo generalised fluctuation-dissipation theorem, the resulting particle motion in the overdamped limit is described by the fractional Langevin equation (FLE)[47,48]

$$\gamma \int_0^t (t-t')^{\alpha-2} \frac{dx(t')}{dt'} dt' = \sqrt{\frac{\gamma k_B \mathscr{T}}{\alpha(\alpha-1)K_\alpha}} \times \xi(t), \qquad (3.4)$$

for $1 < \alpha < 2$. Here $k_B \mathscr{T}$ represents the thermal energy. In this formulation the long-range correlations of the noise are matched by the memory integral over the friction kernel. In terms of the fractional Caputo derivative[49]

$$\frac{d^{2-\alpha}x(t)}{dt^{2-\alpha}} = \frac{1}{\Gamma(\alpha-1)} \int_0^t (t-t')^{\alpha-2} \frac{dx(t')}{dt'} dt' \qquad (3.5)$$

Eq. (3.4) can be rewritten in the compact form

$$\frac{d^{2-\alpha}x(t)}{dt^{2-\alpha}} = \frac{1}{\Gamma(\alpha-1)} \sqrt{\frac{k_B \mathscr{T}}{\gamma\alpha(\alpha-1)K_\alpha}} \times \xi(t), \qquad (3.6)$$

involving the Caputo fractional operator, hence the name fractional Langevin equation.[50] FLE motion is ergodic,

$$\overline{\delta^2(\Delta,T)} \sim \langle x^2(\Delta) \simeq 2K_{2-\alpha}\Delta^{2-\alpha}. \qquad (3.7)$$

Due to the restriction $1 < \alpha < 2$, FLE motion is therefore subdiffusive. As for FBM, the approach to ergodicity is algebraic.[45]

The FLE can be shown to govern the effective dynamics of a tagged particle in a single file[42] or the motion of a monomer in a long polymer chain.[51] The FLE was used to model the internal dynamics of proteins.[52] It occurs naturally for the description of particle motion in a viscoelastic environment,[36] and is related to generalised elastic models[53] as well as hydrodynamic interactions.[54,55] FLE-governed motion was also identified from the motion of individual lipid molecules from large scale simulations of lipid membranes.[39,40] Viscoelasticity controlled subdiffusion was reported for the motion of messenger RNA molecules and chromosomal loci in living *E. coli* cells.[16,56,57] In Ref. [18], the long time motion of lipid granules in living yeast cells was shown to cross over from non-ergodic CTRW motion to viscoelastic-type subdiffusion, consistent with observations in a different strain of yeast cells.[58] In complex fluids, viscoelastic subdiffusion was, *inter alia*, revealed in Refs. [37,38]. Based on microrheology data of endosomes in living cells,[59] stochastic models for active transport in the molecularly crowded cytosol of living cells were recently discussed. These include the viscoelastic nature of crowded fluids[38] in terms of the FLE, and it can be shown that depending on the size of the cargo or the biochemical turnover rate of the motor molecule, normal ($\langle x(t)\rangle \simeq t$) or anomalous ($\langle x(t)\rangle \simeq t^\alpha$ with $0 < \alpha < 1$) transport can be effected.[60] We finally note that the FLE exhibits dynamic transitions with different critical exponents of the driving fractional Gaussian noise for free and processive motion.[47]

3.3. Transient non-ergodicity of FBM & FLE motion

The MSD for both FBM and FLE motion crosses over to a plateau in confinement, for instance, in case of diffusion in an harmonic potential $V(x) = kx^2/2$.[61] In the case of FBM, no temperature is defined, and the value of the stationary plateau is a function of the anomalous diffusion exponent α, $\langle x^2\rangle_{\mathrm{st}} = K_\alpha \Gamma(1 + \alpha)/k^\alpha$.[62] The associated time averaged MSD becomes $\langle \overline{\delta^2}\rangle_{\mathrm{st}} = 2\langle x^2\rangle_{\mathrm{st}}$. Here the factor two between the MSD and the time averaged MSD is due to the definition (1.5), which involves twice the stationary value $\langle x^2\rangle_{\mathrm{st}}$.[63] In contrast to FBM, FLE motion fulfils the fluctuation-dissipation relation, and the MSD relaxes to the unique thermal plateau value $\langle x^2\rangle_{\mathrm{th}} = k_B \mathcal{T}/k$, while the time average converges to $\langle \overline{\delta^2}\rangle_{\mathrm{th}} = 2\langle x^2\rangle_{\mathrm{th}}$.[63]

While for the free FBM and FLE motion ergodic behaviour is found the crossover to the stationary plateau turns out to be transiently non-ergodic. For both FBM and FLE motion the relaxation of the ensemble averaged MSD is exponential. However, for the time averaged MSD the approach is algebraic. For FBM we find[63]

$$\overline{\delta^2(\Delta)} \sim 2\langle x^2\rangle_{\mathrm{st}} - \frac{K_\alpha \Gamma(\alpha + 1)}{k^2}e^{-k\Delta} - \frac{2\alpha(\alpha - 1)K_\alpha}{k^2 \Delta^{2-\alpha}}, \qquad (3.8)$$

and for FLE motion[63]

$$\overline{\delta^2(\Delta)} \sim 2\langle x^2\rangle_{\mathrm{th}}\left(1 - \frac{\gamma}{k\Delta^{2-\alpha}}\right). \qquad (3.9)$$

This transient weak ergodicity breaking may lead to the false assumption that in the analysis of data the process has not yet relaxed, while the corresponding MSD $\langle x^2(t)\rangle$ already reached the plateau. This algebraic return to the ergodic behaviour represented by the plateau reminds of the algebraic approach to ergodicity of the free motion mentioned above. For single particle tracking experiments of submicron tracer beads in a worm-like micellar solution, this behaviour is indeed shown in Fig. 2. In this

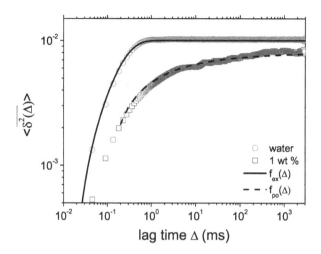

Figure 2. The time averaged MSD of submicron tracer beads in water (circles) and a viscoelastic solution with 1% worm-like micelles (squares).[37] The measurement is based on optical tweezers tracking, so that the initial free motion of the tracer bead eventually becomes confined by the tweezers potential. The time averaged MSD in water relaxes exponentially (full line), while in the worm-like micellar solution we observe the algebraic relaxation of Eq. (3.9), shown by the dashed line.

example the confinement is exerted by the optical tweezers used to track the particle.[37]

3.4. *Transient ageing of FBM & FLE motion*

What happens when the system is initially prepared at time $t = 0$ and we start the measurement at some later time $t_a > 0$, the ageing time? We then define the time averaged MSD as[64,65]

$$\overline{\delta^2(\Delta)} = \frac{1}{T - \Delta} \int_{t_a}^{t_a + T - \Delta} \left[x(t + \Delta) - x(t) \right]^2 dt. \tag{3.10}$$

A Brownian system naturally shows no dependence on t_a. Even though the process is asymptotically ergodic, however, we observe a transient dependence on t_a for processes driven by fGn. In general, for these processes it is found that the time average MSD always contains the two additive terms,[64]

$$\left\langle \overline{\delta^2(\Delta)} \right\rangle = f_{\mathrm{st}}(\Delta) + f_{\mathrm{age}}(\Delta; t_a, T). \tag{3.11}$$

The stationary term depends only on Δ, while the second, ageing term explicitly depends on T and t_a.

Free FLE motion has a stationary term featuring subdiffusion, $f_{st} \simeq \Delta^{2-\alpha}$, and the ageing term decays as $f_{age} \simeq 1/T$ as long as the initial velocity distribution is not thermal. In the limit $t_a \gg T$, we find the ageing time dependence[64]

$$f_{age} \simeq t_a^{-2\alpha}. \tag{3.12}$$

Under confinement FLE motion the term f_{st} has a power-law approach to the thermal plateau value, while again $f_{age} \simeq 1/T$. Interestingly, a different t_a-scaling is followed by the ageing term,[64]

$$f_{age} \simeq t_a^{2\alpha-6}. \tag{3.13}$$

Confined FBM has $f_{age} \simeq 1/T$, however, the ageing term shows the exponential decay[64]

$$f_{age} \sim x_0^2 \exp(-2kt_a). \tag{3.14}$$

4. Subdiffusive continuous time random walks

As discussed in the previous section, FBM and FLE motion reach the ergodic behaviour algebraically, similar to Brownian motion. For sufficiently long measurements, individual trajectories become fully reproducible, and ergodicity is achieved in every single trajectory. Here we introduce a process, for which ergodicity is broken asymptotically, and even for long measurement times T individual trajectories never become reproducible. This process is the well-known Scher-Montroll-Weiss CTRW:[66-68] after each jump a random walker is trapped (immobilised) for some waiting time t before it is allowed to jump again. The waiting times t are independent random variables, that is, CTRWs are renewal processes. Waiting times are distributed identically with the waiting time probability density function $\psi(t)$. The form proposed originally by Scher and Montroll is the power-law[67]

$$\psi(t) \simeq \frac{\tau^\alpha}{t^{1+\alpha}}, \quad 0 < \alpha < 1, \tag{4.1}$$

where τ is a scaling factor with the physical unit of time. With this distribution of waiting times, the process leads to the subdiffusive MSD (2.2).[67,68] Due to the range of α, no characteristic waiting time $\langle t \rangle = \int_0^\infty t\psi(t)dt \to \infty$ exists. This scale-free nature of the CTRW process no longer possesses a time scale that allows one to distinguish a single or few jumps from many jumps. Typically, in a given trajectory longer and longer individual waiting events occur which can become of the order of the measurement time T, no matter how long we run the measurement.

CTRW-type stochastic motion was observed in a wide range of systems, spanning the motion of charge carriers in amorphous semiconductors,[67] the dispersion of tracer chemicals in subsurface aquifers,[69] as well as the motion of tracer beads in cross-linked semiflexible actin gels[70] and of functionalised colloidal particles facing complementarily functionalised surfaces.[71] In living cells, the motion of lipid and insulin granules in the cell cytoplasm[18,19] as well as of protein channels in the plasma membrane[17] follow the law (4.1).

What is the dynamic equation connected to this CTRW process? On the stochastic level, the regular Langevin equation $dx(s)/ds = \xi(s)$ driven by the white Gaussian noise $\xi(s)$ is augmented with a second equation *subordinating* the number of steps s to the real process time t.[72-75] After averaging over the noise, in the diffusion limit we obtain the fractional diffusion equation[28,76]

$$\frac{\partial}{\partial t} P(x, t) = K_\alpha \, _0D_t^{1-\alpha} \frac{\partial^2}{\partial x^2} P(x, t), \qquad (4.2)$$

where we introduced the Riemann-Liouville fractional operator defined by[77,78]

$$_0D_t^{1-\alpha} P(x, t) = \frac{1}{\Gamma(\alpha)} \frac{\partial}{\partial t} \int_0^t \frac{P(x, t')}{(t - t')^{1-\alpha}} dt'. \qquad (4.3)$$

In the limit $\alpha = 1$ we recover the normal diffusion equation (1.1). The fractional diffusion equation (4.2) can equivalently be formulated in terms of the Caputo operator,[78] reading

$$\frac{\partial^\alpha}{\partial t^\alpha} P(x, t) = K_\alpha \frac{\partial^2}{\partial x^2} P(x, t). \qquad (4.4)$$

In Eq. (4.3) we see that the process is dominated by the slowly decaying memory given by the integral over the power-law kernel. In the presence of an external potential, the dynamics is described in terms of the fractional Fokker-Planck equation.[28,79] This fractional Fokker-Planck equation fulfils a generalised form of the Einstein-Stokes relation as well as linear response.[28,79]

The lack of a characteristic waiting time scale effects weak ergodicity breaking,[80,81] and the time averaged MSD becomes[82,83]

$$\left\langle \overline{\delta^2(\Delta)} \right\rangle \sim 2 \frac{K_\alpha}{\Gamma(1 + \alpha)} \frac{\Delta}{T^{1-\alpha}}, \qquad \Delta \ll T, \qquad (4.5)$$

R. Metzler

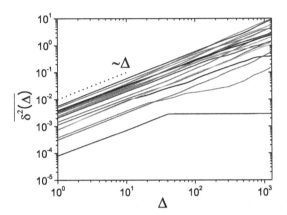

Figure 3.   Individual trajectories of a scale-free, subdiffusive CTRW with $\alpha = 0.5$ exhibit
the linear lag time dependence predicted by Eq. (4.5), with smaller local variations of
the slope. In addition, there is a clear scatter of the amplitudes between individual
trajectories. These features reflect the influence of individual long waiting time events.

which shows a clear disparity with the ensemble averaged MSD (2.2). Despite the anomalous nature of the process, the dependence of the time
averaged MSD (4.5) on the lag time Δ is the same as for Brownian motion.
Only the fact that the amplitude decays as function of the measurement
time T reflects the anomaly: while the process evolves in time, increasingly
longer individual waiting times occur and cause a decay of the effective
diffusivity $\simeq K_\alpha/T^{1-\alpha}$. This behaviour also leads to severe changes in the
interaction of a particle with a reactive surface[84,85] and the exploration of
phase space.[86]

Fig. 3 shows the time averaged MSD for individual realisations of a
subdiffusive CTRW with $\alpha = 0.5$. We notice a distinct scatter of the
amplitudes between the realisations. Moreover, while for most realisations
the predicted linear slope $\left\langle \overline{\delta^2(\Delta)} \right\rangle \simeq \Delta$ is observed, some of the time
traces also show variations in the local slope. Such amplitude scatter and
local slope variations are a common feature in many experiments, compare
Refs. [15,17–19]. We can quantify the amplitude scatter in terms of the
dimensionless ratio $\xi = \overline{\delta^2(\Delta)} \Big/ \left\langle \overline{\delta^2(\Delta)} \right\rangle$. The corresponding distribution
of relative amplitudes, $\phi_\alpha(\xi)$ in the case of the subdiffusive CTRW becomes
a one-sided Lévy stable distribution.[23,82] This distribution for the limit

$T \to \infty$ demonstrates that no matter how long we average the motion of the particle, on the single trajectory level the time averaged MSD of this process always remains a random quantity. In the special case $\alpha = 1/2$ we find the Gaussian form

$$\phi_{1/2}(\xi) = \frac{2}{\pi} \exp\left(-\frac{\xi^2}{\pi}\right). \tag{4.6}$$

Its maximum is at $\xi = 0$, reflecting completely stalled trajectories during the measurement time T. Mobile trajectories with $\xi > 0$ are distributed as a half Gaussian. When α increases towards the Brownian value $\alpha = 1$, a peak emerges at $\xi = 1$. In the Brownian case $\alpha = 1$, ergodicity is restored, and $\phi_1(\xi) = \delta(\xi - 1)$ indicates that for sufficiently long trajectories each realisation is fully reproducible. This behaviour in terms of $\phi(\xi)$ is independent of an external potential,[87,88] due to the fact that the ratio $\overline{\delta^2(\Delta)} \big/ \left\langle \overline{\delta^2(\Delta)} \right\rangle$ is equal to the ratio $n(T)/\langle n(T) \rangle$ of the number of jumps.[65]

Under confinement, for instance, by an harmonic external potential or within a finite domain with reflecting walls, the time averaged MSD of subdiffusive CTRWs does not converge to the thermal plateau of the ensemble averaged MSD. Instead, the time averaged MSD scales like[88,89]

$$\left\langle \overline{\delta^2(\Delta)} \right\rangle \sim \left(\langle x^2 \rangle_B - \langle x \rangle_B^2\right) \frac{2\sin(\pi\alpha)}{(1-\alpha)\pi\alpha} \left(\frac{\Delta}{T}\right)^{1-\alpha} \tag{4.7}$$

for $\Delta \ll T$ and $\Delta \gg (1/[K_\alpha \lambda_1])^{1/\alpha}$. λ_1 represents the lowest non-zero eigenvalue of the Fokker-Planck operator in the confining potential, a measure for the time scale when the particle engages with the confinement. The result (4.7) is universal in so far as only the prefactor depends on the very form of the confining potential $V(x)$. It involves the first and second moments of the Boltzmann distribution, $\langle x^j \rangle_B = \int x^j \exp(-V(x)/[k_B \mathscr{T}])dx/\mathscr{Z}$. The normalisation factor is the partition function $\mathscr{Z} = \int \exp(-V(x)/[k_B \mathscr{T}])dx$. The analysis shows that in this scale free process weak non-ergodicity remains present even in the limit of long measurements.

4.1. *Ageing behaviour of subdiffusive CTRW processes*

CTRW processes with diverging time scale display ageing effects.[90,91] We already saw ageing in the presence of the measurement time T in the time averaged MSD (4.5). Ageing is due to the non-stationarity of the process. Thus, in subdiffusive CTRWs the two-point correlation $\langle x(t_1)x(t_2) \rangle = f(t_1/t_2)$ is not a function of the difference $|t_2 - t_1|$ of the two times but their ratio.[88] This breakdown of stationarity removes the time translation invariance of stationary processes and needs to be taken into consideration in experiments, in which the start of the recording of the trajectories occurs only at some (ageing) time $t_a > 0$ after the original initialisation of the system dynamics at $t = 0$.

For the regular MSD, for sufficiently long ageing times t_a this leads to a crossover from the scaling $\langle x^2(t) \rangle \simeq K_\alpha t / t_a^{1-\alpha}$ in the ageing-dominated regime $t \ll t_a$ to the scaling (2.2) when the system evolves for much longer than the ageing time, $t \gg t_a$.[65,91] In the same situation the time averaged MSD (3.10) behaves much simpler and features the multiplicative, universal correction factor[65]

$$\Lambda_\alpha(t_a/T) = \left(1 + \frac{t_a}{T}\right)^\alpha - \left(\frac{t_a}{T}\right)^\alpha. \tag{4.8}$$

This factor solely depends on the ratio t_a/T of ageing time t_a and measurement time T. Thus, apart from the amplitude, the scaling of the time averaged MSD (3.10) as function of the lag time Δ remains unaffected, an important piece of knowledge when the exact age t_a of the process is not precisely known.[65]

Ageing of a subdiffusive CTRW process gives rise to another remarkable feature. Namely, the probability to observe at least one jump in an aged trajectory of length T decreases algebraically with the ageing time t_a.[65] This property of the *population splitting* of particles into a mobile and a fully immobile fraction has to be taken into account when we want to deduce the anomalous diffusion constant from aged trajectories.[65] We note that also the first passage time behaviour of aged CTRW processes exhibits an explicit dependence on the ageing time t_a. In particular, interesting crossovers between different scaling regimes occur, a fact that may be used to deduce the age t_a of a system from sufficiently long first passage data.[92]

More specifically, in an aged system the start of the measurement at t_a typically finds the system during one of the long waiting time events. It can be shown that the occurrence of the first jump event in this case at the so-called forward (recurrence) waiting time t_1 is distributed according to,[93–95]

$$\psi_1(t_1|t_a) = \frac{\sin(\pi\alpha)}{\pi} \frac{t_a^\alpha}{t_1^\alpha(t_a + t_1)}. \tag{4.9}$$

At long waiting times $t_a \gg t_1$ the distribution of the forward waiting time is thus broader than the regular waiting times t in $\psi(t)$. In an aged CTRW all subsequent jumps then follow the law $\psi(t)$ again. Still, due to the macroscopic memory inherent in CTRW processes,[28] the influence of the ageing time persists until the evolution is much longer than t_a. In a modified CTRW model, in which *every* jump is dominated by the forward waiting time (4.9), the dynamics of the process is significantly slowed down, giving rise to logarithmic time evolutions.[96] These can be connected to single file systems in which each particle separately becomes trapped with a scale-free distribution of trapping times $\psi(t)$.[97]

5. Noisy continuous time random walks

In the subdiffusive CTRW the particle becomes fully immobilised with respect to the co-ordinate $x(t)$ in between successive jump events. For charge carriers in an amorphous semiconductor or for tracer particles firmly stuck to much larger objects or solid surfaces this assumption appears reasonable. However, imagine a submicron tracer particle in a cross-linked network consisting of semi-flexible actin filaments.[70] In this case the particle is stuck in cages for waiting times distributed like the power-law (4.1). The actual trajectory shows distinct fluctuations of approximately constant amplitude around a mean location.[70] This behaviour stems from the thermal nature of the system, that is, the cages in the mesh are typically somewhat larger than the tracer particle and/or the actin filaments making up the mesh are themselves subject to thermal agitation, compare the results of recent simulations of tracer motion in a flexible gel.[98] In the *noisy CTRW* the superimposed noise is combined with the fully immobilised periods of the native CTRW.[99] This model is therefore relevant for the quantitative description of the stochastic particle motion in a large range of systems. In particular, a detailed analysis of recorded data in terms of the noisy CTRW may unveil an underlying power-law waiting time despite the fact that no clear stalling events feature in the measured trajectory.

In the above scenario of approximately constant amplitude noise around the horizontal immobilisation periods in the $x(t)$ diagram, it is a natural choice to add Ornstein-Uhlenbeck noise in the position space to the native subdiffusive CTRW process (see Fig. 4). For the MSD this leads to the additive terms[99]

$$\langle x^2(t) \rangle = \frac{2K_\alpha}{\Gamma(1+\alpha)} t^\alpha + \frac{\eta^2 D}{k} \left(1 - e^{-2kt} \right), \tag{5.1}$$

where the first term represents the contribution of the native CTRW. The second term contains the noise strength $\eta^2 D$ made up of the diffusivity D

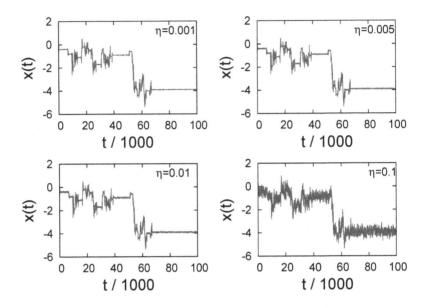

Figure 4. Noisy CTRW process with Ornstein-Uhlenbeck noise with $\alpha = 0.8$, for different amplitudes η of the superimposed Gaussian noise. Increasing η washes out the immobilisation periods of the pure CTRW process.

of physical dimension cm^2/sec and the empirical noise amplitude η. Moreover, k is an inverse time scale governing the relaxation of the Ornstein-Uhlenbeck process to stationarity. The Ornstein-Uhlenbeck component in the MSD (5.1) after the time scale $1/k$ becomes merely an additive constant, whose relative amplitude becomes progressively smaller compared to the first term. The effect on the trajectory itself is displayed in Fig. 4: for increasing noise amplitude the stalling periods of the native CTRW become more and more masked and resemble the experimental trajectories of the submicron tracers in the semi-flexible polymer network.[70]

The associated time averaged MSD becomes[99]

$$\left\langle \overline{\delta^2(\Delta)} \right\rangle \sim \frac{2K_\alpha}{\Gamma(1+\alpha)} \frac{\Delta}{T^{1-\alpha}} + \frac{\eta^2 D}{k} \left(1 - e^{-k\Delta}\right), \qquad (5.2)$$

in absence of ageing ($t_a = 0$). In contrast to the MSD (5.1), the time averaged MSD (5.2) contains the factor $T^{\alpha-1}$ in the first term representing the native CTRW contribution, while the amplitude of the noise in the second term on the right hand side is independent of T. While for small noise amplitude η the observable $\overline{\delta^2}$ will essentially be indistinguishable from the native CTRW, for larger η we observe a distinct crossover behaviour for $\overline{\delta^2}$. Namely, for shorter lag times the time averaged MSD shows contributions from both the native CTRW and the Ornstein-Uhlenbeck noise. Writing $\left\langle \overline{\delta^2} \right\rangle \sim 2D_{\text{app}}\Delta$, for $\Delta \ll 1/k$ we find $D_{\text{app}} \approx K_\alpha T^{\alpha-1}/\Gamma(1+\alpha) + \eta^2 D$. At longer lag times, solely the native CTRW contribution is visible and $D_{\text{app}} \approx K_\alpha T^{\alpha-1}/\Gamma(1+\alpha)$. In between these two regimes, a crossover behaviour is observed. However, when the measurement time T is much longer than the lag time Δ, the Ornstein-Uhlenbeck term is dominant. Again the time average has a clear advantage over the ensemble average, as it reveals additional detail of the behaviour.

A different scenario can also be envisaged.[99] For instance, when the observer is interested in the motion of a tracer inside a living cell and the attachment of the cell to the cover slide in the microscope turns out to be broken, the data will show the additional Brownian noise stemming from the random cell motion superimposed to the anomalous motion with respect to the reference frame of the cell. In that case the MSD reads[99]

$$\langle x^2(t) \rangle = \frac{2K_\alpha}{\Gamma(1+\alpha)}t^\alpha + 2\eta^2 Dt, \qquad (5.3)$$

which exhibits a turnover from the subdiffusive scaling with t^α to the linear Brownian growth $\simeq t$ in the long time limit. The associated time average is always linear,[99]

$$\left\langle \overline{\delta^2(\Delta)} \right\rangle \sim \frac{2K_\alpha}{\Gamma(1+\alpha)}\frac{\Delta}{t^{1-\alpha}} + 2\eta^2 D\Delta. \qquad (5.4)$$

We note that the superposition of Poissonian and non-Poissonian noise was also discussed in a biologically inspired reaction rate model.[100]

6. Ultraslow diffusion of continuous time random walks in an ageing environment

The subdiffusive CTRW process discussed so far is a renewal process. That is, after each step the waiting time τ is randomly chosen from the same PDF $\psi(\tau)$. Physically, this corresponds to an *annealed* environment.[34] More formally, one can view this process as if the random walker carried his own clock around whose random ticks trigger the occurrence of the jumps. As we discussed above, if we want to describe an aged subdiffusive CTRW initiated at $t = 0$ and evolving for the ageing time t_a, the statistics for the occurrence of the first jump in an observation beginning at t_a is modified. The first jump occurs after the forward waiting time t_1, which

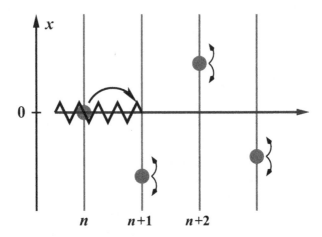

Figure 5.   Sketch of the crack propagation model discussed in the text. The tip of the crack (heavy zig-zag line) propagates from site n to $n+1$ when the vacancy represented by the circle diffuses to the origin ($x = 0$) at point n.

is distributed with the probability density function (4.9). All subsequent waiting times are then again drawn from the standard law (4.1).

Here we consider a different scenario, in which each jump event depends on the present age of the system. Imagine a toy scenario for a rupture model, in the spirit of Zener's famous work on stress relaxation in solids:[101] a crack in a two-dimensional material is propagating along the discrete n axis, as sketched in Fig. 5. The crack is represented by the bold black zig-zag line. As indicated by the arrow, this crack has just propagated from site n to $n+1$. Crack propagation is triggered by the circles (the 'vacancies'), that diffuse along the perpendicular x axis. When the vacancy at site n hit the $x = 0$ line the crack tip was allowed to extend to site $n + 1$. To propagate to site $n + 2$, the vacancy at $n + 1$ has to diffuse to $x = 0$, etc. If the vacancies can only diffuse along a finite interval of length ℓ on the x-axis, they return to $x = 0$ on time scales $\tau_x \simeq \ell^2$. This τ_x then is the average time for the crack propagation from one site to the next, and we will find the crack propagation law $\langle n(t) \rangle \sim t/\tau_x$.

What happens if the length ℓ becomes very large and the vacancies can venture far away? As known from the theory of comb models,[102] the probability density of return to $x = 0$ is of power-law form, proportional to $\tau^{-1-1/2}$. Typically, when the tip of the crack reaches a new site, the vacancy will be away from $x = 0$, and the triggering event for the crack propagation to the next site then corresponds to the forward waiting time t_1 distributed according to Eq. (4.9) with $\alpha = 1/2$. In contrast to the previously discussed renewal ageing CTRWs, however, the next propagation step of the crack tip again occurs with the forward waiting time, characterising the arrival of the next vacancy at $x = 0$, and so forth. In other words, *every* step occurs with the forward waiting time t_1. The probability that the tip arrives at an extremely long forward waiting time t_1 increases considerably.[c] This fact significantly alters the dynamics of the process. Formally, this scenario corresponds to a random walker, which is updated by stationary, site-specific clocks.

If we generalise the CTRW model and consider a process in which every jump occurs with the waiting time PDF (4.1) with general $0 < \alpha < 1$, it can be shown that the crack propagation dynamics is reduced to the much slower logarithmic law[96]

$$\langle n(t) \rangle \sim \frac{\ln(t/t_0)}{\mu}, \tag{6.1}$$

where $\mu = -\Gamma'(\alpha)/\Gamma(\alpha) - \overline{\gamma}$ in terms of the complete Γ function and its derivative Γ', $\overline{\gamma} = 0.5772\ldots$ is Euler's constant, and t_0 is a cutoff time to avoid divergencies at $t = 0$. The counting process $n(t)$ is deterministic in the sense that the relative fluctuations

$$\frac{\sqrt{\langle n^2(t) \rangle - \langle n(t) \rangle^2}}{\langle n(t) \rangle} \simeq \sqrt{\frac{1}{\mu \ln(t/t_0)}}, \tag{6.2}$$

albeit slowly, decrease during the progress of time.[96] For $\alpha > 2$ the process is normal and statistically equivalent to a Poisson update, which is equivalent to the above scenario with finite length ℓ for the vacancy diffusion leading to the linear time dependence $\langle n(t) \rangle \simeq t$. However, similar to our observations above, the case with a finite characteristic update time $\langle \tau \rangle$ but diverging variance of waiting times with $1 < \alpha < 2$ displays the power-law anomaly $\langle n(t) \rangle \simeq t^{\alpha-1}$.[96]

Interestingly, the time average over the time series $n(t)$,[96]

$$\left\langle \overline{n(\Delta)} \right\rangle \sim \frac{1}{\mu T} \ln \left(\frac{T}{t_0} \right) \Delta, \tag{6.3}$$

is linear in the lag time Δ, in analogy to the result (4.5) for the regular renewal subdiffusive CTRW process. The inverse dependence on the measurement time t with the logarithmic correction observed here, in a rough

---

[c]Remember the fact that for long ageing times the probability density function (4.9) of the forward waiting time decays with the power $-\alpha$ and is thus significantly broader as the regular waiting time density (4.1).

way can be viewed as the $\alpha \to 0$ behaviour of the power-law relation in
Eq. (4.5).

Above we constructed the crack propagation model such that the motion
of the tip is fully biased and each step is directed to higher n values. What
if we interpret the update rule for the counting dynamics $n(t)$ as jumps of a
random walk process in real space? To avoid correlations when the random
walker revisits the same spatial point and its next update is governed by the
same clock as during the previous visit, in analogy to the discussion of the
quenched trap model we could include a spatial bias of the random walk.
Alternatively, we could embed the random walk in three dimensions. Due
to the transient nature of this process, revisits are significantly reduced,
and the MSD

$$\langle \mathbf{r}^2(t) \rangle \simeq \ln(t/t_0) \qquad (6.4)$$

of the walker is then proportional to $\langle n(t) \rangle$, while the corresponding time
averaged MSD

$$\left\langle \overline{\delta^2(\Delta)} \right\rangle \simeq \frac{\Delta}{T} \ln(T/t_0) \qquad (6.5)$$

scales like $\left\langle \overline{n(\Delta)} \right\rangle$. Such a random walk process thus exhibits weakly non-
ergodic behaviour.

The random walk process in an ageing environment corresponds to a
non-renewal process in dimension one and two. In dimension three it is
a renewal process, however, here the waiting time distribution (4.1) is replaced by the PDF of the forward (recurrent) waiting time. In other words,
due to the logarithmic nature the process, Eq. (6.1), can be shown to be
governed by the limiting distribution for the product of independent random variables, the log-normal distribution.[96] This approach may thus be
of relevance to a large range of applications in which this distribution is
identified.[103]

In the regular, renewal subdiffusive CTRW ageing affects the statistics
of the first jump, given in terms of the forward waiting time t_1. All subsequent jumps occur with the regular waiting time PDF (4.1). The system
remembers the first step, due to the slowly decaying memory inherent to
the process, seen in the non-local time operator of the associated fractional diffusion equation.[28] Once the process time exceeds the ageing time
significantly, i.e., $T \gg t_a$, the ageing effects are no longer visible.[d] In the

---

[d]This is true for the ensemble averaged MSD as well as for the corresponding time
averaged MSD. In the latter, the ageing depression $\Lambda(t_a/T)$ converges to unity.

non-renewal scenario discussed here the system has a high likelihood to encounter atypically long waiting times at every step and every single step includes ageing. This causes the massive retardation of the motion, giving rise to the emerging logarithmic law. Such time dependencies occur in a large variety of systems, inter alia, the crumpling of paper,[104] compactification of grains,[105] or record statistics.[106] Recently, it was shown that the long time behaviour of a tracer particle in a single file system, in which individual particles repel each other and may stick to a functionalised channel with power-law waiting times, can indeed be described in terms of the logarithmic time dependence derived here within the non-renewal ageing process.[97]

7. Correlated continuous time random walks

What if we do away with the renewal property of the previously discussed CTRW process? One way to include correlations into the CTRW process is to consider a model when successive waiting times are only separated by an incremental change. Physically, this could reflect the motion in a quenched environment, in which locally the motion is dominated by a given mobility with small variations. We could thus imagine that the current waiting τ_i is composed of increments in the form[107-109]

$$\tau_i = \left| \xi_1 + \xi_2 + \ldots + \xi_{i-1} \right|. \tag{7.1}$$

If the ξ_i are distributed according to a Lévy stable law defined in terms of its Fourier transform $\exp\left(-c_\gamma |k|^\gamma\right)$ with $0 < \gamma < 2$, then the process leads to anomalous diffusion governed by Eq. (2.2) with the anomalous diffusion exponent $\alpha = \gamma/(1 + \gamma)$. Its range is $0 < \alpha < 2/3$.[107,108] This model features a stretched exponential mode relaxation $P(k,t) \simeq \exp(-ct^{1/2})$ in the limit $\gamma = 2$, while for for $0 < \gamma < 2$ a power-law form $P(k,t) \simeq t^{-\gamma}$ is obtained.[109] There also exist alternative models to correlated jump processes, see the discussion in Refs. [110,111] and the citations therein.

The absolute value in the law (7.1) implies that the mean waiting time keeps growing with T and diverges in the long time limit. The time averaged MSD[109]

$$\left\langle \overline{\delta^2(\Delta)} \right\rangle \simeq \frac{\Delta}{T^{1-\gamma/(1+\gamma)}} \tag{7.2}$$

shows the weakly non-ergodic behaviour of the correlated CTRW process. It also features ageing effects demonstrated by the temporal decay of the response of the system to a periodic driving force.[109] Individual trajectories

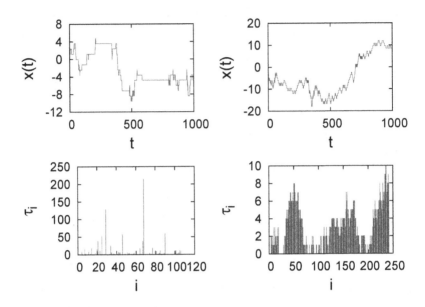

Figure 6.   Trajectory $x(t)$ (**top**) and individual waiting times (**bottom**) in the regular subdiffusive CTRW model with $\alpha = 2/3$ (**left**) and the CTRW model with correlated, Gaussian waiting times[107] with $\gamma = 2$ (**right**). Both cases lead to the same MSD (2.2) with $\alpha = 2/3$.

show a pronounced amplitude scatter.[107]   A graphical comparison of the correlated CTRW with the regular, renewal CTRW is shown in Fig. 6.

A similar trick can be used to correlate subsequent jump lengths. The MSD of this process is then given exactly by[107]

$$\langle x(t)^2 \rangle \simeq \frac{t(t + 1)(2t + 1)\sigma^2}{4}, \tag{7.3}$$

for a Gaussian distribution of jump increments with variance σ^2. This process thus has the cubic long time scaling behaviour $\langle x(t)^2 \rangle \simeq t^3$. The associated time averaged MSD scales quadratically,[107]

$$\left\langle \overline{\delta^2(\Delta)} \right\rangle \simeq \Delta^2 T \tag{7.4}$$

for $\Delta \ll T$. Thus, also this process is weakly non-ergodic.[107]

8. Superdiffusive continuous time random walks and ultra-weak ergodicity breaking

For completeness we also consider superdiffusive renewal CTRW processes. To that end we note that the introduction of a waiting time distribution into a standard random walk process at most leads to a subdiffusive behaviour when the first moment of the waiting time PDF $\psi(\tau)$ diverges. Superdiffusion cannot be achieved within the approach of a generalised waiting time concept. There exist, however, two pathways to extend the CTRW model to superdiffusion.

The first way is to modify the distribution of jump lengths. All CTRW processes considered so far (apart from the case of correlated jump lengths in the preceding section) correspond to the motion on a lattice, or in continuous space with a jump length PDF that possesses a finite variance $\langle \delta x^2 \rangle$ and zero mean $\langle \delta x \rangle$. What if we choose a jump length distribution $\lambda(x)$, for which the variance $\langle \delta x^2 \rangle$ diverges? Consider a Lévy stable form with the asymptotic power-law behaviour $\lambda(x) \simeq 1/|x|^{1+\mu}$ of the jump lengths with the stable index $0 < \mu \le 2$. When the waiting time PDF has finite moments, this process was called a Lévy flight by Benoit Mandelbrot.[112] The divergence of the jump length variance translates into the divergence of the second moment of the PDF $P(x,t)$,[113] and only fractional order moments $\langle |x|^\kappa \rangle$ with $0 < \kappa < \mu$ exist.[28] The trajectory of a Lévy flight is fractal (see below) of Hausdorff dimension μ. A single trajectory therefore never fully covers an embedding space whose dimension is larger than μ. This is particularly relevant in the two-dimensional world, in which effectively most human and animal motion occurs. There exist also several works considering the combination of a diverging characteristic waiting time $\langle \tau \rangle$ with a Lévy stable distribution of jump lengths, either in terms of fractional diffusion equations[114] or via using subordination arguments.[115] Due to its fractality a single Lévy flight trajectory cannot visit all points in space when the stable index μ is smaller than the embedding dimension d. Under confinement to a *finite* area, Lévy flights are ergodic,[116] and the convergence to the ergodic state can be analysed in terms of the apparent fractal dimension or in terms of the first passage dynamics.[117] The divergence of $\langle \delta x^2 \rangle$ as well as the ensuing non-ergodicity of Lévy flights can be rectified by a cutoff in the jump length PDF[118] or by dissipative non-linearities.[119] Such stochastic processes behave like a Lévy flight until the regularisation of the jump length PDF comes into effect.

The alternative approach is to introduce a coupling between jump

lengths and waiting times. In subdiffusive CTRWs described previously the waiting time and jump length PDFs enter in the multiplicative
form $\psi(\delta x, \tau) = \psi(\tau)\lambda(\delta x)$.[120] Introducing a functional dependence between waiting times τ and jump lengths δx, this spatiotemporal coupling
preserves the renewal property of CTRW processes but due to penalisi-
ng long jumps—associating them with long waiting times—yields a finite
MSD.[120,121] The simplest choice is the coupling $\psi(\delta x, \tau) = \frac{1}{2}\psi(\tau)\delta(|\delta x| -
v\tau)$, in which the velocity v is introduced. It bestows a propagating horizon
to the process in the form of two travelling δ peaks with decaying amplitude. For waiting time PDFs $\psi(\tau) \simeq \tau^{-1-\alpha}$ with $1 < \alpha < 2$, in between
these peaks, a Lévy stable distribution is building up.[122] Also Lévy walks
are non-ergodic, albeit in a way, that is different from the above discussed
non-ergodic behaviour. To see this, we first recall that for a waiting time
PDF of the power-law form $\psi(\tau) = \tau^{-1-\alpha}$ their MSD scales[123,124]

$$\langle x^2(t)\rangle \simeq \begin{cases} v^2(1-\alpha)t^2, \, 0 < \alpha < 1 \\ 2K_{3-\alpha}t^{3-\alpha}, \, 1 < \alpha < 2 \end{cases}. \tag{8.1}$$

The associated time averaged MSD in the ballistic phase with $0 < \alpha < 1$
scales like

$$\left\langle \overline{\delta^2(\Delta)} \right\rangle \sim v^2\Delta^2, \tag{8.2}$$

with a higher order correction scaling with $\Delta^2(\Delta/T)^{2-\alpha}$.[125,126] In the
enhanced diffusion phase $1 < \alpha < 2$ the result is[125-127]

$$\left\langle \overline{\delta^2(\Delta)} \right\rangle \sim \frac{2K_{3-\alpha}}{\alpha - 1}\Delta^{3-\alpha}. \tag{8.3}$$

In both the ballistic and enhanced diffusive phases the MSD differs from the
time averaged MSD merely by a factor of $1/|\alpha - 1|$. This phenomenon may
be referred to as ultraweak ergodicity breaking.[127] Note that an analogous
result was obtained by Zumofen and Klafter for Lévy walks with stationary and non-stationary initial conditions,[128] compare the discussion in
Ref. [127]. To leading order, the time averaged MSDs (8.2) and (8.3) do not
exhibit ageing in the sense that the measurement time T does not appear explicitly, in contrast to the corresponding forms for the subdiffusive CTRW
processes discussed above. Further physical properties of Lévy walks, in
particular, the amplitude scatter of the time averaged MSD, are studied
in Refs. [127,129]. Additional recent studies of Lévy walks analyse their
response to an external bias and the power spectral properties.[125-127,129]

    Lévy flights and walks are used as statistical models in many fields,
for example, to quantify blind search processes of animals for sparse food

sources.[130–132] In the science of movement ecology, the so-called Lévy foraging hypothesis has become widely accepted.[131] Recently this model was qualified for human motion behaviour and when different search criteria and external forcing are considered.[133] These stochastic processes also describe the propagation of visible light in disordered optical media[134] and the dynamics of quantum dots.[135] In optical lattices the divergence of the position of single ions were shown to follow Lévy statistics.[136]

9. Scaled Brownian motion

What if we consider a time-dependent diffusion coefficient instead of the x-dependence of heterogeneous diffusion processes (HDPs) considered in the next section? As pointed out by Fuliński already,[137] such experimentally observed variations of the diffusivity[138] may cause weakly non-ergodic behaviour in analogy to the spatial dependence in the HDP process below. For a power-law time dependence of the diffusivity this process is so-called scaled Brownian motion (SBM).[139] Let us start with the Langevin equation

$$\frac{dx(t)}{dt} = \sqrt{2\mathscr{K}(t)} \times \xi(t), \qquad (9.1)$$

where $\xi(t)$ is white Gaussian noise with zero mean. The diffusion coefficient is given by

$$\mathscr{K}(t) = \alpha K_\alpha t^{\alpha-1}, \qquad (9.2)$$

where $0 < \alpha < 2$. This process obviously leads to the MSD (2.2). Concurrently, the time averaged MSD has the exact form[140]

$$\left\langle \overline{\delta^2(\Delta)} \right\rangle = \frac{2K_\alpha T^{1+\alpha}}{(\alpha+1)} \frac{\left[1 - (\Delta/T)^{1+\alpha} - (1 - \Delta/T)^{1+\alpha}\right]}{T - \Delta}. \qquad (9.3)$$

For $\Delta \ll T$, the linear Δ-scaling is recovered,[141]

$$\left\langle \overline{\delta^2(\Delta)} \right\rangle \sim 2K_\alpha \frac{\Delta}{T^{1-\alpha}} \qquad (9.4)$$

in both the sub- and superdiffusive cases. Thus, again we obtain a weakly non-ergodic behaviour given by the disparity between ensemble and time averaged MSD. However, different to the above weakly non-ergodic processes, SBM features fully reproducible trajectories in the long time limit.[140,141] As discussed in Ref. [140] in detail, the time dependent diffusivity $\mathscr{K}(t)$ may appear as a simple and natural choice for the description of anomalous diffusion processes. However, $\mathscr{K}(t)$ actually reflects a time-dependent

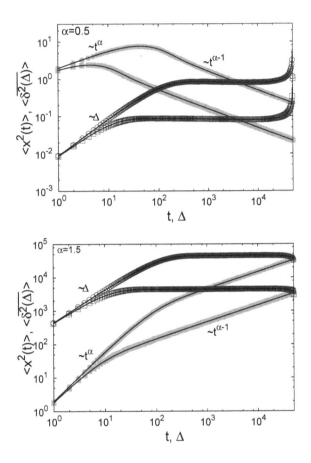

Figure 7.   MSD $\langle x^2(t)\rangle$ (light symbols) and time averaged MSD $\langle \overline{\delta^2(\Delta)}\rangle$ (dark symbols) of SBM with $\alpha = 0.5$ (top) and $\alpha = 1.5$ (bottom). In each case we consider the potential strengths $k^2 = 0.01$ (circles) and $k = 0.1$ (squares). The full lines represent the exact results for the time averaged MSD.[140] The convergence of the corresponding ensemble and time averages at $\Delta = T = 10^5$ can be shown explicitly.

temperature,[137,140] and thus leads to unphysical behaviour in thermalised systems, in particular, when the data are from a confined system, for instance, when the trajectories are measured by optical tweezers methods.[140] This situation is shown in Fig. 7.

However, SBM has a concrete application in granular gases.[142] Granular particles collide inelastically and lose a fraction of their kinetic energy during collisions which transforms into heat stored in internal degrees of freedom. The inelastic nature of inter-particle collisions is the main feature distinguishing a granular gas from a molecular gas, giving rise to many interesting physical properties of dissipative gases. In absence of external forces the gas evolves freely and gradually cools down. During the first stage of its evolution, the granular gas is in the homogeneous cooling state characterised by uniform density and absence of macroscopic fluxes.[143] This may, e.g., be realised in a microgravity environment.[144] Eventually the homogeneous spatial distribution becomes unstable and clusters and vertices develop.[143] In the homogeneous cooling state the behaviour of the granular gas can indeed be captured by SBM.[142]

10. Heterogeneous diffusion processes

Let us now address another seemingly simple anomalous diffusion scenario, a process with a space-dependent diffusivity $K(x)$. Such descriptions were used to model turbulence[24] or diffusion in heterogenous porous media.[145,146] In biological cells, local variations of the diffusion coefficient were indeed recently mapped out.[147] We consider the Langevin equation[148]

$$\frac{dx(t)}{dt} = \sqrt{2K(x)} \times \xi(t), \tag{10.1}$$

where the multiplicative noise $\xi(t)$ is white and Gaussian with zero mean. Using the Stratonovich interpretation this HDP can be shown to be weakly non-ergodic.

Consider the power-law form $K(x) \simeq K_0|x|^\beta$ for the diffusivity, compare Fuliński.[137] The MSD is then given by[148]

$$\langle x^2(t) \rangle = \frac{\Gamma(p+1/2)}{\pi^{1/2}} \left(\frac{2}{p}\right)^{2p} (K_0 t)^p, \tag{10.2}$$

with the exponent $p = 2/(2 - \beta)$. For $\beta < 0$ this process is therefore subdiffusive, while for $0 < \beta < 2$ it is superdiffusive.[148] The time averaged MSD in the limit $\Delta \ll T$ exhibits the linear dependence[148]

$$\left\langle \overline{\delta^2(\Delta)} \right\rangle = \frac{\Gamma(p+1/2)}{\pi^{1/2}} \left(\frac{2}{p}\right)^{2p} K_0^p \frac{\Delta}{T^{1-p}} \tag{10.3}$$

on the lag time, valid for both sub- and superdiffusive regimes. This implies the exact connection $\langle \overline{\delta^2(\Delta)} \rangle = (\Delta/T)^{1-p} \langle x^2(\Delta) \rangle$ with the ensemble averaged MSD.

Interestingly, despite the simplicity of the HDP process we again observe a weakly non-ergodic behaviour. Similar results follow in the case of

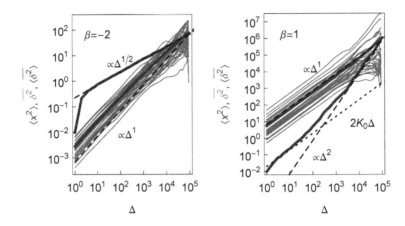

Figure 8. Ensemble and time averaged MSDs for sub- and superdiffusive HDP processes with power-law diffusivity $K(x) \simeq K_0|x|^\beta$, for $T = 10^5$, and $K_0 = 0.01$. Note that in the simulations, to avoid divergence (subdiffusion) or stalling (superdiffusion) at $x = 0$, we respectively use the forms $K(x) = K_0/(x^2 + 1)$ and $K(x) = K_0(|x| + 1)$. Thin lighter lines represent individual traces $\overline{\delta^2(\Delta)}$, thick darker lines refer to the MSD $\langle x^2(t) \rangle$ and the trajectory average $\langle \overline{\delta^2(\Delta)} \rangle$. The expected results (10.2) and (10.3) are shown by dashed lines.

fast (exponential) and slow (logarithmic) variations of the diffusivity $K(x)$ with the particle position x.[149] We note that for the exponential case the square root scaling $\langle \overline{\delta^2(\Delta)} \rangle \simeq \Delta^{1/2}$ was observed.[149] In the context of imaged diffusion in cells the HDP process with power-law x-dependence of $K(x)$ was also generalised to two dimensions.[150] Fig. 8 shows the behaviour of the MSD for the free HDP. We note that the ageing dynamics of HDPs with power-law diffusivities closely resemble those of subdiffusive CTRWs.[151] However, the MSD and time averaged MSD for *confined* HDPs rather saturate to a plateau[151] and thus behave similar to confined FBM and confined Brownian motion, while no saturation is observed for confined CTRWs.[88,89]

11. Sinai diffusion and ultraslow continuous time random walks

In the theory of stochastic processes, the logarithmic time evolution has a prominent representative, namely, Sinai diffusion.[152] In this special case of Temkin's model,[153] the random walker moves in the quenched energy landscape created by a seed random walk. Thus, locally the walker experiences a force of the same amplitude, randomly to the left or the right. The walker can become trapped significantly when the bias in a number of adjacent sites point in direction of the walker's current location. To get to a distance x from its starting point the particle needs to cross an energy barrier of the typical order $\sqrt{x}$, corresponding to an activation time scale $\tau \simeq \tau_1 \exp(c\sqrt{x})$, where τ_1 is a fundamental time scale and c a dimensional constant. The typical distance covered by the walker during time t then scales according to the ultraslow, logarithmic law $x^2 \simeq \ln^4(t/\tau_1)$.[34] Referring to Ref. [154] for further explanations, we quote the result for the time averaged MSD

$$\left\langle \widetilde{\delta^2(\Delta)} \right\rangle \simeq \frac{3721}{17080} \ln^4(T) \frac{\Delta}{T} = \widetilde{\langle x^2(T) \rangle} \frac{549}{854} \frac{\Delta}{T}, \tag{11.1}$$

where the tilde denotes the disorder average. Interestingly, also here the time averaged MSD increases linearly with the lag time and exhibits a strong sensitivity to the measurement time. A generalisation of the Sinai model with strongly correlated disorder was reported recently.[155]

In terms of a renewal CTRW ultraslow processes can be established by using a waiting time PDF of the form $\psi(t) \simeq 1/(t \log^{1+\gamma} t)$,[154,156–158] which is normalised but does not possess finite moments of any power $\langle \tau^q \rangle$ with $q > 0$. It produces an MSD of the form

$$\langle x^2(t) \rangle \simeq \log^\gamma t, \tag{11.2}$$

i.e., for $\gamma = 4$ the MSD scales identically to that of the Sinai diffusion. The weakly non-ergodic behaviour of ultraslow CTRWs is analogous to Eq. (11.1) for Sinai diffusion, apart from the general exponent γ and the prefactor,

$$\left\langle \widetilde{\delta^2(\Delta)} \right\rangle \sim \langle x^2(T) \rangle \times \frac{\Delta}{T}. \tag{11.3}$$

The time averaged MSD, the localisation of the diffusion particle, as well as the ergodic properties of both Sinai and ultraslow CTRW diffusion are analysed in Ref. [154], discussing some of the fundamental differences between time averages recorded in annealed versus quenched environments.

12. Interacting many particle systems: single file diffusion and generalisation

Many-body interactions represent one of the fundamental problems in physics and chemistry.[159] A generic model combining many-body interactions with the stochastic motion of the individual particles in the system is single file motion, in which diffusing particles with hard core, excluded volume interactions move in one dimension. The resulting collisions between the particles in the single file severely alter the Brownian law performed by any of the particles in the case when no excluded volume interactions were present. As shown by Harris in 1965, the motion of a labelled tracer particle in a single file of particles is characterised by the square-root scaling

$$\langle x^2(t) \rangle \simeq K_{1/2} t^{1/2} \tag{12.1}$$

of the MSD.[41] Out of the various theoretical and numerical approaches we mention in particular the harmonisation method, showing that after integrating out all other particle co-ordinates the motion of the labelled tracer particle is described by a fractional Langevin equation[42] (see Refs. [42,97] for additional References).

It was recently studied how disorder affects the dynamics of an interacting many-body system such as a single file. Based on a physical scenario it was demonstrates that a single, non-interacting particle in the presence of annealed disorder performs anomalous diffusion characterised by the power-law form $\langle x^2(t) \rangle \simeq t^\alpha$ of the mean squared displacement with $0 < \alpha < 1$, corresponding to the subdiffusive CTRW from section 4. In the presence of the additional many-body effects in the single file, however, one observes the emergence of ultraslow, logarithmic motion of the form[97]

$$\langle x^2(t) \rangle \simeq \log^{1/2} t \tag{12.2}$$

for the labelled tracer particle. Mathematically, it can be shown that this result can be understood from subordination of the many-body Harris result (12.1) in absence of the disorder to the dynamics of a counting process in an ageing system[96] (see section 6), leading to the universal 1/2 exponent of the ensuing logarithmic law (12.2). The remarkable slowing-down in this system, due to the conspiration of disorder and strong particle-particle interactions, represents a generic effect for ultraslow dynamics in low-dimensional many-body systems.

13. Diffusion on fractals

An important approach to the description of porous or crowded media is the percolation model.[112,160] In site percolation each point on a lattice is occupied with probability p and remains empty with probability $1 - p$. At the critical occupation probability $p = p_c$ ($p_c \approx 0.59\ldots$ in two[161] and

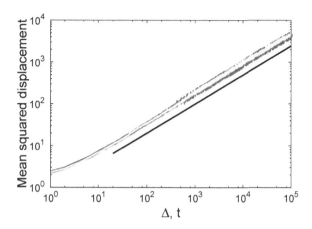

Figure 9. MSD (thicker darker curve) and time averaged MSD (thinner lighter curve) of a random walk on the infinite critical percolation cluster. Both MSD and time averaged MSD perfectly overlap, i.e., diffusion on a fractal is stationary and ergodic. The straight line shows the expected slope $\alpha = 0.697$ to guide the eye. Data provided by Y. Meroz, corresponding to those used in Ref. [168].

$p_c \approx 0.31\ldots$ in three[162] dimensions for a square and cubic lattice, respectively) the correlation length of the system diverges and an infinite cluster is formed. The percolation cluster then has a fractal dimension $d_f = 91/48 \approx 1.896\ldots$ in two[163] and $d_f \approx 2.52\ldots$ in three dimensions.[164] A random walker placed on the fractal, incipient infinite cluster allowed to move between nearest neighbour occupied sites performs anomalous diffusion with an anomalous diffusion exponent $\alpha = 2/d_w$ related to the walk exponent d_w, which is larger than d_f.[165,166] According to the Alexander Orbach conjecture $d_w = \frac{3}{2}d_f$,[166] which is close to experimentally observed values, compare Ref. [167]. Note that when the averaged motion of random walkers placed on *all* clusters, a different scaling exponent characterises the MSD, for more details see Ref. [31].

Results for both the two-dimensional MSD $\langle \mathbf{r}^2(t) \rangle$ and time averaged MSD $\overline{\delta^2}$ for the motion on the infinite cluster are shown in Fig. 9. Both overlap perfectly, corroborating the ergodicity of this anomalous diffusion process.[168] The straight line in Fig. 9 indicates the expected slope to guide the eye. In Ref. [169], the non-Gaussian nature of the diffusion on the critical percolation cluster is analysed. Fractal percolation clusters are often used for simulations of free diffusive processes[170] as well as facilitated diffusion processes[146] in the crowded cytoplasm of living biological cells. A

fractal support was also diagnosed to be superimposed onto the subdiffusive CTRW motion for the diffusion of potassium channels in the plasma
membrane of living human cells in Ref. [17]. It is important to note that
when we consider the motion on all clusters the motion is a forteriori no
longer ergodic: a walker moving on a finite, disconnected cluster cannot
explore the entire phase space.[31]

While the increments of random walk processes on fractal structures are
stationary[32] and the infinite percolation cluster simulations of Ref. [168]
indicate that diffusion on fractals is ergodic, this point needs further investigation, in particular, for different types of fractals. A second open
question is what happens in the presence of a topological bias, for instance,
a bias away from the backbones[160] of a diffusion cluster. In that case at
least transient non-ergodicity would be expected.

14. Conclusions

Single particle tracking is increasingly becoming a standard tool to study
the motion of tracer particles in systems such as complex fluids or even
living biological cells. Concurrently, single particle traces are evaluated in
large scale computer simulations, for instance, to detect inhomogeneous
motion in a population of simulated particles. To evaluate the garnered
time series one typically uses the time averaged MSD. As we showed here,
when the motion of the particle is anomalous, care has to be taken to
evaluate the results in a physically meaningful way. Due to the occurrence
of transient or asymptotic weak ergodicity breaking, one cannot simply
compare the results for the time averages with the known behaviour of the
corresponding ensemble averages.

Apart from the processes discussed herein, non-ergodic behaviour also
occurs in other stochastic processes, including the ultraweakly non-ergodic
Lévy walks[125–128] where the disparity between ensemble and time averaged
MSDs only amounts to a constant factor. Diffusion on random, fractal
percolation clusters was shown to be ergodic.[171] We also note that in some
systems combinations of stochastic processes have to be applied to capture
the observed data.[17–19,172–174]

The diagnosis of a given data set for the exact underlying stochastic process[29–32] requires the analysis of several complementary quantities. We mention the amplitude scatter statistics,[46] increment autocorrelati-
ons,[23,40] higher order moments,[175,176] mean maximal excursion methods,[175]
p-variation,[177,178] and the analysis of the distribution of the apparent dif-

fusivity.[179]

Acknowledgements

The author acknowledges funding from the Academy of Finland within the Finland Distinguished Professor scheme.

References

1. R. Brown, Phil. Mag. **4**, 161, 1828.
2. T. L. Carus, *De rerum natura (50 BCE)*, *On the nature of things*, Harvard University Press, Cambridge, Massachusetts, 1975.
3. J. Ingenhousz, *Nouvelles expériences et observations sur divers objets de physique*, T. Barrois le jeune, Paris, 1785.
4. A. Fick, Ann. Phys. (Leipzig) **170**, 50 (1855).
5. A. Einstein, Ann. d. Physik **17**, 549 (1905).
6. K. Pearson, Nature **72**, 294 (1905).
7. M. von Smoluchowsky, Ann. Phys. (Leipzig) **21**, 756 (1906).
8. P. Langevin, C. R. Acad. Sci. Paris **146**, 530 (1908).
9. W. Sutherland, Philos. Mag. **9**, 781 (1905).
10. T. N. Thiele, Vidensk. Selsk. Skr. 5. Rk., naturvid. og mat. Afd. **12**, 381 (1880).
11. J. Perrin, C. R. Acad. Sci. Paris **146**, 967 (1908).
12. I. Nordlund, Z. Phys. Chem. **87**, 40 (1914).
13. E. Kappler, Ann. d. Phys. (Leipzig) **11**, 233 (1931).
14. C. Bräuchle, D. C. Lamb, and J. Michaelis, *Single Particle Tracking and Single Molecule Energy Transfer* (Wiley-VCH, Weinheim, Germany, 2012); X. S. Xie, P. J. Choi, G.-W. Li, N. K. Lee, and G. Lia, Annu. Rev. Biophys. **37**, 417 (2008).
15. I. Golding and E. C. Cox, Phys. Rev. Lett. **96**, 098102 (2006).
16. S. C. Weber, A. J. Spakowitz, and J. A. Theriot, Phys. Rev. Lett. **104**, 238102 (2010).
17. A. V. Weigel, B. Simon, M. M. Tamkun, and D. Krapf, Proc. Nat. Acad. Sci. USA **108**, 6438 (2011).
18. J.-H. Jeon, V. Tejedor, S. Burov, E. Barkai, C. Selhuber-Unkel, K. Berg-Sørensen, L. Oddershede, and R. Metzler, Phys. Rev. Lett. **106**, 048103 (2011).
19. S. M. A. Tabei, S. Burov, H. Y. Kim, A. Kuznetsov, T. Huynh, J. Jureller, L. H. Philipson, A. R. Dinner, and N. F. Scherer, Proc. Natl. Acad. Sci. USA **110**, 4911 (2013).
20. G. Seisenberger, M. U. Ried, T. Endreß, H. Büning, M. Hallek, and C. Bräuchle, Science **294**, 1929 (2001).
21. A. Caspi, R. Granek, and M. Elbaum, Phys. Rev. Lett. **85**, 5655 (2000).
22. N. Gal and D. Weihs, Phys. Rev. E **81**, 020903(R) (2010).

23. S. Burov, J.-II. Jeon, R. Metzler, and E. Barkai, Phys. Chem. Chem. Phys. **13**, 1800 (2011).
24. L. F. Richardson, Proc. Roy. Soc. London, Ser. A **110**, 709 (1926); A. S. Monin and A. M. Yaglom, *Statistical Fluid Mechanics* (MIT Press, Cambdridge MA, 1971).
25. A. Smekal, Physikal. Zeitschr. **27**, 383 (1926).
26. H. Freundlich and D. Krüger, Trans. Faraday Soc. **31**, 906 (1935).
27. R. O. Herzog and A. Polotzky, Zeitschr. f. Physikal. Chemie – Stochiometrie und Verwandtschaftslehre **87**, 449 (1914).
28. R. Metzler and J. Klafter, Phys. Rep. **339**, 1 (2000); J. Phys. A **37**, R161 (2004).
29. M. J. Saxton and K. Jacobson, Annu. Rev. Biophys. Biomol. Struct. **26**, 373 (1997).
30. E. Barkai, Y. Garini, and R. Metzler, Physics Today **65**(8), 29 (2012).
31. F. Höfling and T. Franosch, Rep. Prog. Phys. **76**, 046602 (2013).
32. I. M. Sokolov, Soft Matter **8**, 9043 (2012).
33. R. Metzler, J.-H. Jeon, A. G. Cherstvy, and E. Barkai, Phys. Chem. Chem. Phys. **16**, 24128 (2014).
34. J.-P. Bouchaud, J. Phys. (Paris) I **2**, 1705 (1992).
35. W. T. Coffey and Yu. P. Kalmykov, *The Langevin Equation: With Applications to Stochastic Problems in Physics, Chemistry and Electrical Engineering* (World Scientific, Singapore, 2012).
36. I. Goychuk, Phys. Rev. E **80**, 046125 (2009); Adv. Chem. Phys. **150**, 187 (2012).
37. J.-H. Jeon, N. Leijnse, L. B. Oddershede, and R. Metzler, New J. Phys. **15**, 045011 (2013).
38. J. Szymanski and M. Weiss, Phys. Rev. Lett. **103**, 038102 (2009).
39. G. R. Kneller, K. Baczynski, and M. Pasienkewicz-Gierula, J. Chem. Phys. **135**, 141105 (2011); M. Javanainen, H. Hammaren, L. Monticelli, J.-H. Jeon, R. Metzler, and I. Vattulainen, Faraday Discussions **161**, 397 (2013).
40. J.-H. Jeon, H. Martinez-Seara Monne, M. Javanainen, and R. Metzler, Phys. Rev. Lett. **109**, 188103 (2012).
41. T. E. Harris, J. Appl. Prob. **2**(2), 323 (1965).
42. L. Lizana, T. Ambjörnsson, A. Taloni, E. Barkai, and M. A. Lomholt, Phys. Rev. E **81**, 051118 (2010).
43. A. N. Kolmogorov, Dokl. Acad. Sci. USSR **26**, 115 (1940).
44. B. B. Mandelbrot and J. W. van Ness, SIAM Rev. **1**, 422 (1968).
45. W. Deng and E. Barkai, Phys. Rev. E **79**, 011112 (2009).
46. J.-H. Jeon and R. Metzler, J. Phys. A **43**, 252001 (2010).
47. S. Burov and E. Barkai, Phys. Rev. Lett. **100**, 070601 (2008); Phys. Rev. E **78**, 031112 (2008).
48. R. Zwanzig, *Nonequilibrium Statistical Mechanics* (Oxford University Press, Oxford, UK, 2001).
49. F. Mainardi, *Fractional calculus and waves in linear viscoelasticity* (Imperial College Press, London, 2010).
50. E. Lutz, Phys. Rev. E **64**, 051106 (2001).

51. D. Panja, J. Stat. Mech. **L02001** (2010); **P06011** (2010).
52. S. C. Kou and X. S. Xie, Phys. Rev. Lett. **93**, 180603 (2004).
53. A. Taloni, A. V. Chechkin, and J. Klafter, Phys. Rev. Lett. **104**, 160602 (2010).
54. T. Franosch, M. Grimm, M. Belushkin, F. M. Mor, G. Foffi, L. Forro, and S. Jeney, Nature **478**, 7367 (2011); M. Grimm, S. Jeney, and T. Franosch, Soft Matt. **7**, 2076 (2011).
55. D. S. Grebenkov, M. Vahabi, E. Bertseva, L. Forro, and S. Jeney, Phys. Rev. E **88**, 040701 (2013); D. S. Grebenkov and M. Vahabi, Phys. Rev. E **89**, 012130 (2014).
56. K. Burnecki, E. Kepten, J. Janczura, I. Bronshtein, Y. Garini, and A. Weron, Biophys. J. **103**, 1839 (2012); E. Kepten, I. Bronshtein, and Y. Garini, Phys. Rev. E **83**, 041919 (2011).
57. M. Magdziarz, A. Weron, K. Burnecki and J. Klafter, Phys. Rev. Lett. **103**, 180602 (2009).
58. M. A. Taylor, J. Janousek, V. Daria, J. Knittel, B. Hage, H.-A. Bachor, and W. P. Bowen, Nature Phot. **7**, 229 (2013).
59. D. Robert, T. H. Nguyen, F. Gallet, and C. Wilhelm, PLoS ONE **4**, e10046 (2010).
60. I. Goychuk, V. O. Kharchenko, and R. Metzler, PLoS ONE **9**, e91700 (2014); Phys. Chem. Chem. Phys. **16**, 16524 (2014).
61. J.-H. Jeon and R. Metzler, Phys. Rev. E **81**, 021103 (2010).
62. O. Yu. Sliusarenko, V. Yu. Gonchar, A. V. Chechkin, I. M. Sokolov, and R. Metzler, Phys. Rev. E **81**, 041119 (2010).
63. J.-H. Jeon and R. Metzler, Phys. Rev. E **85**, 021147 (2012).
64. J. Kursawe, J. H. P. Schulz, and R. Metzler, Phys. Rev. E **88**, 062124 (2013).
65. J. H. P. Schulz, E. Barkai, and R. Metzler, Phys. Rev. Lett. **110**, 020602 (2013); Phys. Rev. X **4**, 011028 (2014).
66. E. W. Montroll and G. H. Weiss, J. Math. Phys. **6**, 167 (1965).
67. H. Scher and E. W. Montroll, Phys. Rev. B **12**, 2455 (1975).
68. B. D. Hughes, *Random Walks and Random Environments, Volume 1: Random Walks* (Oxford University Press, Oxford, 1995).
69. H. Scher, G. Margolin, R. Metzler, J. Klafter, and B. Berkowitz, Geophys. Res. Lett. **29**, 1061 (2002).
70. I. Y. Wong, M. L. Gardel, D. R. Reichman, E. R. Weeks, M. T. Valentine, A. R. Bausch, and D. A. Weitz, Phys. Rev. Lett. **92**, 178101 (2004).
71. Q. Xu, L. Feng, R. Sha, N. C. Seeman, and P. M. Chaikin, Phys. Rev. Lett. **106**, 228102 (2011).
72. H. C. Fogedby, Phys. Rev. E **50**, 1657 (1994).
73. A. Baule and R. Friedrich, Phys. Rev. E **71**, 026101 (2005).
74. W. Feller, *An introduction to probability theory and its application* (Wiley, New York, NY, 1970).
75. M. Magdziarz and A. Weron, Phys. Rev. E **75**, 016708 (2007).
76. W. R. Schneider and W. Wyss, J. Math. Phys. **30**, 134 (1989).
77. K. B. Oldham and J. Spanier, *The fractional calculus* (Academic Press, New York, NY, 1974).

78. I. Podlubny, *Fractional Differential Equations* (Academic Press, New York, NY, 1998).
79. R. Metzler, E. Barkai, and J. Klafter, Phys. Rev. Lett. **82**, 3563 (1999); E. Barkai, R. Metzler, and J. Klafter, Phys. Rev. E **61**, 132 (2000); R. Metzler, E. Barkai, and J. Klafter, Europhys. Lett. **46**, 431 (1999).
80. J.-P. Bouchaud, J. Phys. I **2**, 1705 (1992).
81. G. Bel and E. Barkai, Phys. Rev. Lett. **94**, 240602 (2005).
82. Y. He, S. Burov, R. Metzler, and E. Barkai, Phys. Rev. Lett. **101**, 058101 (2008).
83. A. Lubelski, I. M. Sokolov, and J. Klafter, Phys. Rev. Lett. **100**, 250602 (2008).
84. M. A. Lomholt, I. M. Zaid, and R. Metzler, Phys. Rev. Lett. **98**, 200603 (2007); I. M. Zaid, M. A. Lomholt, and R. Metzler, Biophys. J. **97**, 710 (2009).
85. M. J. Skaug, A. M. Lacasta, L. Ramirez-Piscina, J. M. Sancho, K. Lindenberg, and D. K. Schwartz, Soft Matter **10**, 753 (2014); M. Khoury, A. M. Lacasta, J. M. Sancho, and K. Lindenberg, Phys. Rev. Lett. **106**, 090602 (2011).
86. G. Bel and E. Barkai, J. Phys. Cond. Mat. **17**, S4287 (2005); Phys. Rev. Lett. **94**, 240602 (2005); A. Rebenshtok and E. Barkai, J. Stat. Phys. **133**, 565 (2008); Phys. Rev. Lett. **99**, 210601 (2007).
87. I. M. Sokolov, E. Heinsalu, P. Hänggi and I. Goychuk, Europhys. Lett. **86**, 30009 (2009).
88. S. Burov, R. Metzler, and E. Barkai, Proc. Natl. Acad. Sci. USA **107**, 13228 (2010).
89. T. Neusius, I. M. Sokolov, and J. C. Smith, Phys. Rev. E **80**, 011109 (2009).
90. C. Monthus and J.-P. Bouchaud, J. Phys. A **29**, 3847 (1996).
91. E. Barkai and Y. C. Cheng, J. Chem. Phys. **118**, 6167 (2003).
92. H. Krüsemann, A. Godec, and R. Metzler, Phys. Rev. E **89**, 040101(R) (2014).
93. E. B. Dynkin, Izv. Akad. Nauk. SSSR Ser. Math. **19**, 247 (1955); Selected Translations Math. Stat. Prob. **1**, 171 (1961).
94. C. Godrèche and J. M. Luck, J. Stat. Phys. **104**, 489 (2001); E. Barkai and Y.-C. Cheng, J. Chem. Phys. **118**, 6167 (2003); E. Barkai, Phys. Rev. Lett. **90**, 104101 (2003).
95. T. Koren, M. A. Lomholt, A. V. Chechkin, J. Klafter, and R. Metzler, Phys. Rev. Lett. **99**, 160602 (2007).
96. M. A. Lomholt, L. Lizana, R. Metzler, and T. Ambjörnsson, Phys. Rev. Lett. **110**, 208301 (2013).
97. L. P. Sanders, M. A. Lomholt, L. Lizana, K. Fogelmark, R. Metzler, and T. Ambjörnsson, New J. Phys. (at press); E-print arXiv:1311.3790.
98. A. Godec, M. Bauer, and R. Metzler, New J. Phys. **16**, 092002 (2014).
99. J.-H. Jeon, E. Barkai, and R. Metzler, J. Chem. Phys. **139**, 121916 (2013).
100. S. Eule and R. Friedrich, Phys. Rev. E **87**, 032162 (2013).
101. C. Zener, Elasticity and anelasticity of metals (University of Chicago Press, Chicago, IL, 1948); Phys. Rev. **52**, 230 (1937); *ibid.* **53**, 90 (1938).

102. G. H. Weiss and S. Havlin, Physica A **13**, 474 (1986); S. Havlin, J. E. Kiefer, and G. H. Weiss, Phys. Rev. A **36**, 1403 (1987).
103. E. Limpert, W. A. Stahel, and M. Abbt, BioScience **51**, 341 (2001).
104. K. Matan, R. B. Williams, T. A. Witten, and S. R. Nagel, Phys. Rev. Lett. **88**, 076101 (2002).
105. P. Richard, M. Nicodemi, R. Delannay, P Ribière, and D. Bideau, Nature Mat. **4**, 121 (2005).
106. B. Schmittmann and R.K.P. Zia, Am. J. Phys. **67**, 1269 (1999).
107. V. Tejedor and R. Metzler, J. Phys. A **43**, 082002 (2010).
108. M. Magdziarz, R. Metzler, W. Szczotka, and P. Zebrowski, Phys. Rev. E **85**, 051103 (2012).
109. M. Magdziarz, R. Metzler, W. Szczotka, and P. Zebrowski, J. Stat. Mech. P04010 (2012).
110. A. V. Chechkin, M. Hofmann, and I. M. Sokolov, Phys. Rev. E **80**, 031112 (2009).
111. J. H. P. Schulz, A. V. Chechkin, and R. Metzler, J. Phys. A. **46**, 475001 (2013).
112. B. B. Mandelbrot, *The fractal geometry of nature* (W. H. Freeman, New York, NY, 1982).
113. H. C. Fogedby, Phys. Rev. Lett. **73**, 2517 (1994); S. Jespersen, R. Metzler, and H. C. Fogedby, Phys. Rev. E **59**, 2736 (1999).
114. A. I. Saichev and G. M. Zaslavsly, Chaos **7**, 753 (1997); B. J. West and T. F. Nonnenmacher, Phys. Lett. A **278**, 255 (2001); Y. Luchko and R. Gorenflo, Fract. Calc. Appl. Anal. **1**, 63 (1998); R. Metzler and T. F. Nonnenmacher, Chem. Phys. **284**, 67 (2002).
115. M. Magdziarz and A. Weron, Phys. Rev. E **75**, 056702 (2007).
116. A. Weron and M. Magdziarz, Phys. Rev. Lett. **105**, 260603 (2010).
117. M. Vahabai, J. H. P. Schulz, B. Shokri, and R. Metzler, Phys. Rev. E **87**, 042136 (2013).
118. R. N. Mantegna and H. E. Stanley, Phys. Rev. Lett. **73**, 2946 (1994).
119. A. V. Chechkin, V. Yu. Gonchar, J. Klafter, and R. Metzler, Phys. Rev. E **72**, 010101(R) (2005).
120. J. Klafter, A. Blumen, and M. F. Shlesinger, Phys. Rev. A **35**, 3081 (1987).
121. M. F. Shlesinger, J. Klafter, and Y. M. Wong, J. Stat. Phys. **27**, 499 (1982).
122. J. Klafter and G. Zumofen, Phys. Rev. E **49**, 4873 (1994).
123. G. Zumofen, J. Klafter, and A. Blumen, Chem. Phys. **146**, 433 (1990).
124. J. Masoliver, K. Lindenberg, and G. H. Weiss, Physica A **157**, 891 (1989).
125. D. Froemberg and E. Barkai, Phys. Rev. E **87**, 030104(R) (2013); Phys. Rev. E **88**, 024101 (2013).
126. D. Froemberg and E. Barkai, Euro. Phys. J. B **86**, 331 (2013).
127. A. Godec and R. Metzler, Phys. Rev. Lett. **110**, 020603 (2013); Phys. Rev. E **88**, 012116 (2013).
128. G. Zumofen and J. Klafter, Physica D **69**, 436 (1993).
129. M. Niemann, H. Kantz, and E. Barkai, Phys. Rev. Lett. **110**, 140603 (2013); G. Margolin and E. Barkai, J. Stat. Phys. **122**, 137 (2006).
130. D. W. Sims, M. J. Witt, A. J. Richardson, E. J. Southall, and J. D. Metcalfe,

Proc. Biol. Sci. **273**, 1195 (2006); N. E. Humphries et al, Nature **475**, 1066 (2010); G. M. Viswanathan et al, Nature **381**, 413 (1996); A. M. Edwards et al, Nature **449**, 1044 (2007); N. E. Humphries, H. Weimerskirch, N. Queiroz, E. J. Shouthall, and D. W. Sims, Proc. Natl. Acad. Sci. USA **109**, 7169 (2012).

131. G. E. Viswanathan, M. G. E. da Luz, E. P. Raposo, and H. E. Stanley, *The physica of foraging: an introduction to random searches and biological encounters* (Cambridge University Press, Cambridge, UK, 2011).

132. M. F. Shlesinger and J. Klafter, in *On growth and form*, edited by H. E. Stanley and N. Ostrowsky (Martinus Neijhoff, Dordrecht, The Netherlands, 1986); G. M. Viswanathan et al, Nature **401**, 911 (1999); M. A. Lomholt, T. Koren, R. Metzler, and J. Klatfer, Proc. Natl. Acad. Sci USA **105**, 1055 (2008); F. Bartumeus, J. Catalan, U. L. Fulco, M. L. Lyra, and G. M. Viswanathan, Phys. Rev. Lett. **88**, 097901 (2002).

133. M. C. González, C. A. Hidalgo, and A.-L. Barabási, Nature **453**, 779 (2008); C. M. Song, T. Koren, P. Wang, and A.-L. Barabási, Nature Phys. **6**, 818 (2010); V. V. Palyulin, A. V. Chechkin, and R. Metzler, Proc. Natl. Acad. Sci. USA **111**, 2931 (2014).

134. P. Barthelemy, J. Bertolotti, and D. S. Wiersma, Nature **453**, 495 (2008).

135. F. D. Stefani, J. P. Hoogenboom, and E. Barkai, Phys. Today **62**(2), 34 (2009); G. Margolin and E. Barkai, Phys. Rev. Lett. **94**, 080601 (2005); X. Brokmann, J.-P. Hermier, G. Messin, P. Desbiolles, J.-P. Bouchaud, and M. Dahan, Phys. Rev. Lett. **90**, 120601 (2003).

136. H. Katori, S. Schlipf, and H. Walther, Phys. Rev. Lett. **79**, 2221 (1997).

137. A. Fuliński, Phys. Rev. E **83**, 061140 (2011); J. Chem. Phys. **138**, 021101 (2013).

138. M. Platani, I. Goldberg, A. I. Lamond, and J. R. Swedlow, Nature Cell Biol. **4**, 502 (2002).

139. S. C. Lim and S. V. Muniandy, Phys. Rev. E **66**, 021114 (2002).

140. J.-H. Jeon, A. V. Chechkin, and R. Metzler, Phys. Chem. Chem. Phys. **16**, 15811 (2014).

141. F. Thiel and I. M. Sokolov, Phys. Rev. E **89**, 012115 (2014).

142. A. Bodrova, A. V. Chechkin, and R. Metzler (unpublished).

143. N. V. Brilliantov and T. Pöschel, *Kinetic theory of Granular Gases*, (Oxford University Press, Oxford, UK, 2004).

144. Y. Grasselli, G. Bossis, and G. Goutallier, Europhys. Lett. **86**, 60007 (2009).

145. M. Dentz, P. Gouze, A. Russian, J. Dweik, and F. Delay, Adv. Water Res. **49**, 13 (2012).

146. C. Loverdo et al., Phys. Rev. Lett. **102**, 188101 (2009).

147. T. Kühn, T. O. Ihalainen, J. Hyväluoma, N. Dross, S. F. Willman, J. Langowski, M. Vihinen-Ranta, and J. Timonen, PLoS One **6**, e22962 (2011).

148. A. G. Cherstvy, A. V. Chechkin, and R. Metzler, New J. Phys. **15**, 083039 (2013).

149. A. G. Cherstvy and R. Metzler, Phys. Chem. Chem. Phys. **15**, 20220 (2013).

150. A. V. Cherstvy, A. V. Chechkin, and R. Metzler, Soft Matter **10**, 1591 (2014).

151. A. G. Cherstvy, A. V. Chechkin, and R. Metzler, J. Phys. A (at press).
152. Ya. G. Sinai, Theory Prob. Appl. **27**, 256 (1982).
153. D. E. Temkin, Sov. Math. Dokl. **13**, 1172 (1972).
154. A. Godec, A. V. Chechkin, E. Barkai, H. Kantz, and R. Metzler, J. Phys. A (at press); E-print arXiv:1406.6199.
155. G. Oshanin, A. Rosso, and G. Schehr, Phys. Rev. Lett. **110**, 100602 (2013).
156. S. Denisov and H. Kantz, Phys. Rev. E **83**, 041132 (2011); S. I. Denisov, S. B. Yuste, Yu. S. Bystrik, H. Kantz, and K. Lindenberg, Phys. Rev. E **84**, 061143 (2011).
157. S. Havlin and G. H. Weiss, J. Stat. Phys. **58**, 1267 (1990).
158. J. Dräger and J. Klafter, Phys. Rev. Lett. **84**, 5998 (2000).
159. L. Samaj and Z. Bajnok, *Introduction to the Statistical Physics of Integrable Many-body Systems* (Cambridge University Press, Cambridge, UK, 2013); G. Stefanucci and R. van Leeuwen, *Nonequilibrium Many-Body Theory of Quantum Systems: A Modern Introduction* (Cambridge University Press, Cambridge, UK, 2013).
160. J. Feder, Fractals (Plenum Press, New York, NY, 1988).
161. J. Lykke Jacobsen, J. Phys. A **47**, 135001 (2014).
162. J. Wang, Z. Zhou, W. Zhang, T. M. Garoni, and Y. Deng, Phys. Rev. E **87**, 052107 (2013).
163. B. Nienhuis, J. Stat. Phys. **34**, 731 (1984).
164. C. D. Lorenz and R. M. Ziff, Phys. Rev. E **57**, 230 (1998).
165. Y. Gefen, A. Aharony, and S. Alexander, Phys. Rev. Lett. **50**, 77 (1983).
166. S. Alexander and R. Orbach, J. Physique (Paris) **43**, L625 (1982).
167. A. Klemm and R. Kimmich, Phys. Rev. E **55**, 4413 (1997); A. Klemm, R. Metzler, and R. Kimmich, Phys. Rev. E **65**, 021112 (2002).
168. Y. Meroz, I. M. Sokolov, and J. Klafter, Phys. Rev. E **81**, 010101(R) (2010).
169. M. Spanner, F. Höfling, G. E. Schröder-Turk, K. Mecke, and T. Franosch, J. Phys. Cond. Mat. **23**, 234120 (2011).
170. M. Saxton, Biophys. J. **103**, 2411 (2012); **72**, 1744 (1997).
171. Y. Meroz, I. Eliazar, and J. Klafter, J. Phys. A **42**, 434012 (2009); Y. Meroz, I. M. Sokolov, and J. Klafter, Phys. Rev. Lett. **110**, 090601 (2013).
172. T. Akimoto, E. Yamamoto, K. Yasuoka, Y. Hirano, and M. Yasui, Phys. Rev. Lett. **107**, 178103 (2011).
173. J.-H. Jeon, E. Barkai, and R. Metzler, J. Chem. Phys. **139**, 121916 (2013).
174. S. Eule and R. Friedrich, Phys. Rev. E **87**, 032162 (2013).
175. V. Tejedor, O. Bénichou, R. Voituriez, R. Jungmann, F. Simmel, C. Selhuber-Unkel, L. Oddershede, and R. Metzler, Biophys. J. **98**, 1364 (2010).
176. D. Ernst, J. Kohler, and M. Weiss, Phys. Chem. Chem. Phys. **16**, 7686 (2014).
177. M. Magdziarz, A. Weron, K. Burnecki, and J. Klafter, Phys. Rev. Lett. **103**, 180602 (2009).
178. K. Burnecki, E. Kepten, J. Janczura, I. Bronshtein, Y. Garini, and A. Weron, Biophys. J. **103**, 1839 (2012).
179. M. Bauer, R. Valiullin, G. Radons, and J. Kaerger, J. Chem. Phys. **135**, 144118 (2011).

Chapter 4

Kinetics of Pattern Formation: Mesoscopic and Atomistic Modelling

Héléna Zapolsky*

GPM, UMR 6634, University of Rouen, av. de l'Université, BP-12 76801
Saint-Etienne du Rouvray, France,
helena.zapolsky@univ-rouen.fr

The phenomenon of pattern formation is universal in nature and involves complex non-equilibrium processes that are highly important for both fundamental research and applied materials systems as well as for theoretical biology. Last decades, computational modelling provided an important tool for understanding pattern formation dynamics and self-organisation in many systems. In this chapter the brief overview of different methods of modelling of pattern formation phenomena has been done. The linking between mesoscopic and atomistic approaches is established. It is shown that some general conclusions about pattern formation can be done from the modelling of microstructure evolution in alloys at these two scales.

Contents

---

*helena.zapolsky@univ-rouen.fr

1. Introduction

For a long time, pattern formation has fascinated the human mind. We find patterns everywhere. Examples can be found in almost every field of today's scientific interest, ranging from coherent pattern formation in physical and chemical systems[1,4] to the various morphogenetic problems in biology[5,7] and many examples of self-organisation in the universe (solar granulation, sand dunes on Mars, vortex formations such as the Large Red Spot on Jupiter and the structure of planetary rings, etc.).[8,9] At present, self-assembly in the periodic structures at the nanoscale is becoming increasingly important for the fabrication of novel supramolecular structures, with applications in the fields of nanobiotechnology and nanomedicine.[10,11] In spite of a large diversity of patterns some of them are observed more frequently than others. If we look around us it is easy to remark that the more frequently observed patterns in nature are the stripe and the honeycomb (or hexagonal) structures, the animal coat patterns being of this type (see Fig. 1).

Figure 1.   Patterns in animal coats.

The phenomena of pattern formation in physical experiments was evocated for the first time by Bénard in 1900. It was shown that in a plane horizontal layer of a fluid heated from below above a critical threshold a regular pattern of convection cells, known as Bénard cells, with different symmetries (rolls, stripes and hexagons) appear. Rayleigh has explained this phenomenon in 1916. It was established that the competition between the buoyancy forces, a friction and hence the diffusion between hotter and cooler regions of the fluid is responsible for the appearance of convection cells (see Fig. 2).

The next important step in understanding pattern formation was done by Alain Turing in 1952 in his famous article *The Chemical Basis of Morphogenesis*.[13] It was demonstrated for the first time how a simple model

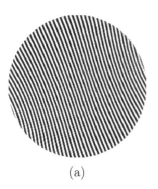
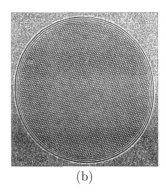

(a)   (b)

Figure 2.   Two convection patterns.[12]

system of coupled reaction-diffusion equations could give rise to spatial patterns in chemical concentrations through a process of chemical instability. The best-known Turing patterns are composed of stripes or simple hexagonal arrangements of spots. It was shown also that Turing instability might play a major role in the generation of skin patterns in a number of animals.[14] Some years later the hexagonal patterns were also observed in type II superconductors (see Fig. 3). In these systems the magnetic field can penetrate a sample in a tube-like configuration. The tubes of magnetic flux are referred to as (superconducting) vortices arranged in the lattice with hexagonal symmetry. In this case system is in a mixed state where the superconducting and non-superconducting phases coexist. In this mixed state the interaction between vortices can be considered as a competition between a repulsion between two neighbouring tubes of magnetic flux which can be associated with two wires with the opposite direction of current and an attraction that appears from the minimization of free energy (decreasing of the interface between superconducting and non-superconducting phases). This competition between the two opposite forces results in the formation of a hexagonal pattern. From all these examples the main question appears: Why in so different phenomena that happen at different time and space scales the more stable patterns are already stripe and hexagonal structures?

In the last decades, computational modelling provided an important tool for understanding pattern formation dynamics and self-organization in many systems.[15,16] In the last 25 years, significant progress has been done by using molecular dynamics (MD) simulations of biological macromolecules.[17] However, in spite of very big advances in different fields of

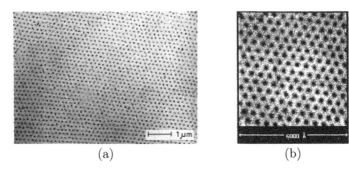

(a)   (b)

Figure 3.   Vortex structure in the type II superconductors. (a) First picture of vortices obtained by Essman and Traeuble by decoration in Pb-4%I at 1.1K.[18] (b) Vortex structure in NbSe$_2$ obtained by scanning tunnelling microscopy.[19]

science that study pattern formation, there is no general answer to the previous question.

In this chapter the modelling of pattern formation during phase transformation at mesoscopic and atomic scale is considered. It is shown that using simulations some progress in understanding the pattern formation can be done. The chapter is organized as follows: in Section 2 we report the theoretical description of pattern formation in reaction diffusion reactions and convection processes. Then in Section 3 we discuss the microscopic description of the microstructure evolution in solid-solid reaction. We show that the introduction of the anisotropic long-range potential induces very rich morphology in a microstructure in alloys. In the last section we discuss a mesoscopic description of the same phenomena. Some examples of the simulated microstructures are provided. In this section the link between the microscopic and mesoscopic descriptions of the microstructure in solids has been done. We conclude with some brief remarks about universality and predictability in the description of pattern formation by different approaches.

2. Patterns in reaction-diffusion and convection phenomena

2.1. Reaction-diffusion equation

In the quest for understanding biological growth, it was Alan Turing who first demonstrated how a simple model system of coupled reaction-diffusion equations could give rise to spatial patterns in chemical concentrations through a process of chemical instability. In the article *The Chemical Basis of Morphogenesis*[13] he proposed an explanation of how the patterns of animals like leopards, jaguars and zebras are determined. Turing asserted that the patterns could arise as a result of instabilities in the diffusion of morphogenetic chemicals in the animals skin during the embryonic stage of development.

To introduce some general aspect of this approach, let us consider the simplest case of a system in which the chemical reactions lead to the synthesis or to the destruction of two chemical species A and B. We will note, eventually, the concentration of the species A and B at time t and at position $\mathbf{r}$, $u_A(\mathbf{r}, t)$ and $u_B(\mathbf{r}, t)$, respectively. As was proposed in Ref. [13], the temporal evolution of these concentrations is controlled by reaction-diffusion equations:

$$\frac{\partial u_A(\mathbf{r}, t)}{\partial t} = f(u_A(\mathbf{r}, t), u_B(\mathbf{r}, t)) + D_A \nabla^2 u_A(\mathbf{r}, t), \qquad (2.1)$$

$$\frac{\partial u_B(\mathbf{r}, t)}{\partial t} = g(u_A(\mathbf{r}, t), u_B(\mathbf{r}, t)) + D_B \nabla^2 u_B(\mathbf{r}, t). \qquad (2.2)$$

The first terms in Eq.(2.1) and Eq.(2.2), $f(u_A(\mathbf{r}, t), u_B(\mathbf{r}, t))$ and $g(u_A(\mathbf{r}, t), u_B(\mathbf{r}, t))$, are proportional to the rate of the appearance (or of the destruction) of chemical species A and B, respectively. The second terms take into account the phenomenon of diffusion and the coefficients D_A and D_B are the diffusion coefficients of the species A and B. In the homogeneous system concentrations of two chemical species are constants, $u_A(\mathbf{r}, t) = u_A^0$ and $u_B(\mathbf{r}, t) = u_B^0$, and the system is in a stationary state. Then, to explore the stability of this stationary state with respect to the fluctuations of concentration of species A and B, $\Delta u_A(\mathbf{r}, t) = u_A^0 - u_A(\mathbf{r}, t)$ and $\Delta u_B(\mathbf{r}, t) = u_B^0 - u_B(\mathbf{r}, t)$, we should linearise the initial system of kinetic equations in the vicinity of this state and then analyse the solution of this linearised system. In a given case, after a linearisation of the system of equations (2.1) and (2.2), we search the solution in the next form: $\Delta u_A(\mathbf{r}, t), \Delta u_B(\mathbf{r}, t) \propto e^{(\lambda t + i\mathbf{k}\mathbf{r})}$. Here $\mathbf{k}$ is the wave vector and λ can be found in the characteristic equation of the system of linearised kinetic equations. It is easy to see that if λ is positive, the fluctuations with wave vector $\mathbf{k}$ will grow in time. The solution of the characteristic equation also gives the relationship between λ and $\mathbf{k}$. The relationship connecting λ and $\mathbf{k}$ has a special name - "dispersion relationship"- and characterizes the chemical properties of the system. For the present case this dispersion relationship is:

$$\lambda(k^2) = -\frac{B}{2} \pm \frac{\sqrt{B^2 - 4C}}{2}, \qquad (2.3)$$

where

$$B = k^2(D_A + D_B) - Tr\ A,$$

$$C = det\ A + k^4 D_A D_B - k^2(\frac{\partial f}{\partial u_A} D_B + \frac{\partial g}{\partial u_B} D_A).$$

Here $det\ A$ and $Tr\ A$ is a determinant and a trace of the matrix A which is:

$$A = \begin{pmatrix} \frac{\partial f}{\partial u_A} - \lambda - k^2 D_A & \frac{\partial f}{\partial u_A} \\ \frac{\partial g}{\partial u_B} & \frac{\partial g}{\partial u_A} - \lambda - k^2 D_B \end{pmatrix}.$$

In order to identify the conditions under which the system becomes unstable with respect to the fluctuations with wave vector k_0 or some group of wave vectors, some requirements on the terms that go into Eq. (2.3) should be satisfied. It was shown by Turing that the system becomes unstable with respect to the fluctuations if:

$$\frac{\partial f}{\partial u_A} + \frac{\partial g}{\partial u_B} < 0, \tag{2.4a}$$

$$\frac{\partial f}{\partial u_A}\frac{\partial g}{\partial u_B} + \frac{\partial f}{\partial u_B}\frac{\partial g}{\partial u_A} > 0, \tag{2.4b}$$

$$\frac{\partial f}{\partial u_A} D_B + \frac{\partial f}{\partial u_B} D_A > 0, \tag{2.4c}$$

$$(\frac{\partial f}{\partial u_A} D_B + \frac{\partial f}{\partial u_B} D_A)^2/4D_A D_B > \frac{\partial f}{\partial u_A}\frac{\partial g}{\partial u_B} + \frac{\partial f}{\partial u_B}\frac{\partial g}{\partial u_A}. \tag{2.4d}$$

Let us remark that λ is a quadratic function of $\mathbf{k}$. The general behaviour of λ as function of $\mathbf{k}$ is shown in Fig. 4. As it follows from this figure, there is a window of k^2 values, limited by k_1^2 and k_2^2, for which the systems are unstable. However, the fluctuations with the wave vector corresponding to the maximum value of the function $\lambda(k^2)$ will grow faster then the other ones.

The values of k_1^2 and k_2^2 can be found as a root of the next expression:

$$k_{1,2}^2 = \frac{\frac{\partial f}{\partial u_A} D_B + \frac{\partial g}{\partial u_B} D_A}{2D_A D_B} \pm \frac{\sqrt{(\frac{\partial f}{\partial u_A} D_B + \frac{\partial f}{\partial u_B} D_A)^2 - 4D_A D_B det A}}{2D_A D_B}. \tag{2.5}$$

Using the inequalities Eqs. (2.4a) and (2.4c) Turing gave an interpretation of some phenomena observed in chemistry and biology. As it follows from the previous conditions (Eqs. (2.4a)-(2.4d)) set forth, the variables

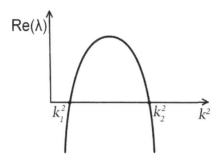

Figure 4.   Schematic representation of the function $\lambda(k^2)$.

$u_A(\mathbf{r}, t)$ and $u_B(\mathbf{r}, t)$ have the following properties: if $\frac{\partial f}{\partial u_A} > 0$ and $\frac{\partial g}{\partial u_B} > 0$ and at the same time $D_B > D_A$, then chemical species A induces its own production while B inhibits its own production. This fact is responsible for the introduction of "activator-inhibitor" term in reaction-diffusion phenomena. It is very important to remark that there is a coupling between the diffusion coefficients and the characteristic times of the activator and the inhibitor. In order for that system to become unstable, the inhibitor must diffuse faster than the activator and the activator has a shorter lifetime than inhibitor. This situation is seen in many biological systems and can explain the pattering of animal coats. The solution of Eqs. (2.1) and (2.2) in two dimensional (2D) case gives only two types of patterns: a stripe and a honeycomb.

Murray, in his book *Mathematical Biology*,[20] suggested also that a single mechanism could be responsible for generating all of the common patterns observed in animal coats. He considered the production and diffusion of melanin, a pigment that affects skin, eye, and hair color in humans and other mammals. Solving a reaction-diffusion system of equations in a 2D case on finite domain with no-flux boundary conditions he described the way in which two different chemical products react and are propagated on the skin, one that colors the skin, and one that does not colour it; or more precisely, one that stimulates the production of melanin (colouring the skin) and one that inhibits this production. What is remarkable is that the equations show that the different patterns of coat depend only on the size and form of the region where they are developed. Stated in another way, the same basic equation explains all of the patterns. But then, why do a tiger and a leopard have different patterns given that their bodies are similar? This is because the formation of the patterns would not be produced at the same moment during the growth of the embryo. In the first instance, the embryo would still be small and in the other, it would be at a much bigger stage.

More precisely, the equations show that no pattern is formed if the embryo is very small, that a striped pattern is formed if the embryo is a little bigger, a spotted pattern if it is bigger yet, and no pattern at all if it is too big. This is why a mouse and an elephant would not have a pattern. What is more, as to comparable surfaces, the form of the surface makes a difference. Thus, if one considers a certain surface large enough to permit the formation of spots, and that one gives it a long, cylindrical form (like a tail) without altering its total area, then the spots are transformed into stripes. In this way, a unique system of differential equations seems to govern all the coat patterns that one finds in nature. The same type of equations also permits one to explain the patterns of the wings of butterflies as well as certain colored patterns of exotic fish.

To conclude this part let us remark that the reaction diffusion system of equations can be also interpreted differently. The right part of the system of Eqs. (2.4a)-(2.4d) can be seen as a first derivative of the free energy of a system with respect to the concetration of A and B atoms. It can be divided into two contributions: the local (first term) part and non-local (second term) one. The local part corresponds to the local production (or destruction) of atom of species α ($\alpha = A$ or B). The non-local part (gradient term) can be interpreted as an interaction between the atoms A and B. In this case, the function $\lambda(k^2)$ can be interpreted as a Fourier transform of this interaction. This function is an isotropic one minimum function. For our further discussions, we will keep in mind that if the kinetic equation contains the interaction, which can be described by the isotropic one-mode potential, the stable solutions in 2D case are stripe or honeycomb periodic structures.

2.2. *Swift-Hohenberg equation and pattern formation in convection processes*

Let us turn back to the Rayleigh-Bénard convection problem and consider (at first, qualitatively) what happens with an arbitrary small spatially non-uniform perturbation of a trivial state upon a slight rise over the convection threshold. In the simplest version of this problem a layer of liquid confined between two infinite stress free plates is heated uniformly from below. If the difference in temperature ΔT between two plates is small, then energy is transported by molecular conduction. As ΔT is increased and achieved the critical value ΔT_c, the conduction state loses stability. The rising and falling regions eventually form cellular structures known as convection rolls. The characteristic roll size is about the distance between the plates. The acceleration of the hot regions by buoyancy forces is opposed by a friction arising from the fluid viscosity and also by the diffusion of heat between warmer and cooler regions of the fluid. The Rayleigh number R is a dimensionless constant that characterises the ratio between all these forces:

$$R = \frac{\alpha g \Delta T d^3}{\kappa \upsilon}, \tag{2.6}$$

where α is the fluid's coefficient of thermal expansion, g is the gravitational acceleration, d is the distance between the plates, κ is the fluid's thermal diffusivity and ν is the fluid's kinematic viscosity. For a given system the liquid loses stability at some critical value of R_c.

Swift and Hohenberg[21] proposed a simple model in 1976 to describe a spatiotemporal pattern formation in these convectional processes.[21] In their famous work on fluctuations near the onset of the Rayleigh-Bénard convection, Swift and Hohenberg derived an order parameter equation for the temperature and fluid velocity dynamics of the convection. In this model the order parameter is associated to the scalar non-conserved variable $\psi(x, y, t)$ which is the vertical component of the velocity of fluid perpendicular to the plates surface. The temporal evolution of this variable is described by the Landau-Khalatnikov relaxation equation:[22]

$$\frac{\partial \psi}{\partial t} = -\frac{\delta F}{\delta \psi}, \tag{2.7}$$

where F is the free energy energy!free of the system. For the free energy of the heterogeneous 2D system it was proposed to write in the gradient form:

$$F = \int \int \left\{ \frac{\psi}{2} \left[\nabla^2 + k_0^2 \right] \psi + \frac{\psi^4}{4} \right\} dx dy, \tag{2.8}$$

where $|\mathbf{k}_0|$ and ϵ are constants of the model. It can be easily shown that the minimum of function F falls at $\mathbf{k}_0$. It means that the heterogeneities with a wave vector $\mathbf{k}_0$ will grow faster than other ones. The constant ϵ is the control parameter related to the Rayleigh number and it is proportional to deviations of the Rayleigh number from the critical value at which the convective instability occurs:

$$\varepsilon = \frac{R - R_c}{R_c} = \frac{T - T_c}{T_c}. \tag{2.9}$$

The total free energy described by Eq. (2.8) consists of the bulk free energy and interfacial energy (gradient terms). As in the previous paragraph, we can see that the free energy of the system contains two contributions: local, related to the free energy of the homogeneous state (the terms in powers of ψ) and non-local which is described by the gradient term. Again this non-local term of the free energy can be interpreted as an interaction between the cells in convection processes.

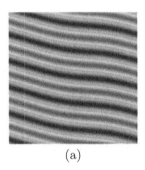   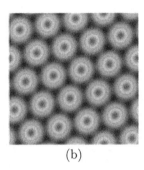

(a)   (b)

Figure 5.   Morphology of patterns obtained by the numerical solution of Eq. (2.10) with the next set of parameters: $|\mathbf{k}_0| = 1$ and (a) $\varepsilon = 0.1$, (b) $\varepsilon = 0.3$.

To conclude this part we would like to underline that the application of the Eq. (2.10) is not limited to the original physical problem and it thus represents a generic model for the spatiotemporal dynamics of spatially extended patterns.

By substituting Eq. (2.8) into the kinetic equation (2.7) we obtain:

$$\frac{\partial \psi}{\partial t} = \left[\varepsilon - \nabla^2 + k_0^2 \right]^2 \psi - \psi^3. \tag{2.10}$$

If $\epsilon < 0$ (R$< R_c$) the stationary solution of Eq. (2.10) corresponds to the homogeneous state with $\psi(x, y, t) = constant$. This solution represents the homogeneous immobile liquid where the energy is transported by molecular conduction. However, if $\epsilon > 0$ (R$> R_c$) the homogeneous state becomes unstable and periodic convective patterns appear. The stationary solution of Eq. (2.10) in this case corresponds to the periodic structure with wave vectors $\mathbf{k}_i$ which satisfy the condition. For the different absolute values of ϵ in the 2D case the two solutions can be stable for the small absolute value of ϵ the period structure corresponds to the periodic "honeycomb" structure with a period equal to $a_0 = \frac{4\pi}{k_0\sqrt{3}}$ with increasing of ϵ the stripe structure becomes stable with the lattice parameter equal to $a_0 = \frac{2\pi}{k_0}$. To illustrate these two cases the numerical solution of (2.10) has been obtained for two values of ϵ. The final structure obtained in these simulations is shown in Fig. 5. The colour scheme used to visualise the value of the order parameter ψ is as follows. If the value of ψ is close to 1, then this site is shown in white. If ψ is close to 0 then this site is shown in black. In the intermediate cases where $0 < \psi < 1$ (mostly at the interface between the two phases), the probabilities are shown as a shade of grey.

3. Modelling of microstructure evolution in solids

3.1. *Stability of a homogeneous system to phase separation and ordering*

At approximately the same time parallel to Turing work, Russian physicist Lev Landau[22,23] developed the phenomenological theory of phase transitions. He also analysed the criteria where a physical system becomes unstable with respect to fluctuations. In our case we will consider that these fluctuations are related to variation of concentration in a system.

In this approach, near the temperature of a second order phase transition T_c the functional of the free energy F of a system can be expanded in the Taylor series with respect to the amplitude of the fluctuations $\delta c(\mathbf{r})$:

$$F = F_{hom} + \sum_{\mathbf{r}} A(\mathbf{r})\delta c(\mathbf{r}) + \frac{1}{2}\sum_{\mathbf{r},\mathbf{r}'} B(\mathbf{r},\mathbf{r}')\delta c(\mathbf{r})\delta c(\mathbf{r}') +$$

$$\frac{1}{3!}\sum_{\mathbf{r},\mathbf{r}',\mathbf{r}''} C(\mathbf{r},\mathbf{r}',\mathbf{r}'')\delta c(\mathbf{r})\delta c(\mathbf{r}')\delta c(\mathbf{r}'') + ..., \qquad (3.1)$$

where F_{hom} is a free energy of a homogeneous state ($\delta c(\mathbf{r}) = 0$), $A(\mathbf{r})$, $B(\mathbf{r},\mathbf{r}')$ and $C(\mathbf{r},\mathbf{r}',\mathbf{r}'')$ are the expansion coefficients. The summations in Eq. (3.1) are carried out over all crystal lattice sites. As well as in the disordered state all sites are crystallographically equivalents and we consider the system with conserved number of atoms ($\sum_{\mathbf{r}} \delta c(\mathbf{r}) = 0$). Near the temperature of a second order phase transition where the amplitudes of fluctuations are small we can consider only the first non-vanishing term in (3.1):

$$\Delta F = F - F_{hom} = \frac{1}{2}\sum_{\mathbf{r},\mathbf{r}'} B(\mathbf{r},\mathbf{r}')\delta c(\mathbf{r})\delta c(\mathbf{r}')\delta c(\mathbf{r}''), \qquad (3.2)$$

where ΔF is a variation of free energy due to the heterogeneity $\delta c(\mathbf{r})$. Using the concentration wave approximation proposed by Khachaturyan[24] the fluctuations of the concentration of different atomic species in a system can be written as:

$$\delta c(\mathbf{r}) = \sum_{\mathbf{k}}{}' Q(\mathbf{k}) e^{i\mathbf{k}\mathbf{r}}, \qquad (3.3)$$

where $Q(\mathbf{k})$ is the concentration wave amplitude, the prime in Eq. (3.3) indicates that $\mathbf{k}=0$ is excluded from the summation. Using Eq. (3.3) the Fourier transform of Eq. (3.2) is:

$$\Delta F = \frac{N}{2} \sum_{\mathbf{k}} b(\mathbf{k}) \left| Q(\mathbf{k}) \right) |^2, \qquad (3.4)$$

where

$$b(\mathbf{k}) = \sum_{\mathbf{r}} B(\mathbf{r}) e^{i\mathbf{k}\mathbf{r}}. \qquad (3.5)$$

The summation in (3.4) is carried out over all N wave vectors within the first Brillouin zone allowed by the periodic boundary conditions and N is the total number of crystal lattice sites.

From the definitions (3.4) and (3.5) the function $b(\mathbf{k}; T, c, p)$ is the thermodynamic function of the disordered state and consequently depends on temperature, concentration and pressure. At a high temperature, where the homogeneous state is stable, all coefficients $b(\mathbf{k}; T, c, p)$ in Eq. (3.4) are positive. With decreasing temperature, the minimum of the function $b(\mathbf{k}_0; T, c, p)$, corresponding to $\mathbf{k} = \mathbf{k}_0$, becomes zero and the homogeneous state loses its stability. In the future we will consider a system at a given temperature, concentration and pressure, and to simplify our equations, we will drop the thermodynamic parameters T, c and p from the definition of the function $b(\mathbf{k})$.

To determine the form of the function $b(\mathbf{k})$ let us consider the chemical free energy F_{chem} of a binary system in a mean-field approximation. For simplicity let us consider a binary solid solution. Then, an arbitrary state of this system can be described by the atomic occupation density functions $p_A(\mathbf{r}, t)$ and $p_B(\mathbf{r}, t)$ that are the occupation probabilities of finding A or B atoms, respectively, at lattice site $\mathbf{r}$ at a given time t. In a binary substitutional solution the functions, $p_A(\mathbf{r}, t)$ and $p_B(\mathbf{r}, t)$ are not independent. They are related by the identity, $p_A(\mathbf{r}, t) + p_B(\mathbf{r}, t) = 1$. Therefore, the atomic configuration of the system can be fully described by just one density function. For certainty, we assume that this function is $p_A(\mathbf{r}, t) = p(\mathbf{r}, t)$. In equilibrium the function $p(\mathbf{r})$ is independent on time. The chemical free energy of the binary system can be interpolated by the equation for the mean-field free energy:[25]

$$F_{chem} = \frac{1}{2}\sum_{\mathbf{r}}\sum_{\mathbf{r}'} W(\mathbf{r} - \mathbf{r}')p(\mathbf{r})p(\mathbf{r}') + k_B T\sum_{\mathbf{r}}[p(\mathbf{r})\ln p(\mathbf{r}) +$$

$$(1 - p(\mathbf{r}))\ln(1 - p(\mathbf{r}))], \tag{3.6}$$

where $W(\mathbf{r} - \mathbf{r}') = W_{AA}(\mathbf{r} - \mathbf{r}') + W_{BB}(\mathbf{r} - \mathbf{r}') - 2W_{AB}(\mathbf{r} - \mathbf{r}')$ is a mixing energy, $W_{\alpha\beta}(\mathbf{r} - \mathbf{r}')$ are the pairwise interaction energies between a pair of atoms, α and β (=A or B,), at lattice site $\mathbf{r}$ and $\mathbf{r}'$. It is convenient to rewrite the first term in (3.6) using Fourier space representation:

$$F_{chem} = \frac{1}{2N}\sum_{\mathbf{k}} V(\mathbf{k})\,|\tilde{p}(\mathbf{k}))|^2 + k_B T\sum_{\mathbf{r}}[p(\mathbf{r})\ln p(\mathbf{r}) +$$

$$(1 - p(\mathbf{r}))\ln(1 - p(\mathbf{r}))], \tag{3.7}$$

where $\tilde{p}(\mathbf{k}) = \sum_{\mathbf{k}} p(\mathbf{r})e^{-i\mathbf{kr}}$ and $V(\mathbf{k}) = \sum_{\mathbf{r}} W(\mathbf{r})e^{-i\mathbf{kr}}$ are the Fourier transforms of $p(\mathbf{r})$. Summation over $\mathbf{r}$ in Eq. (3.7) is carried out over all N lattice sites. Summation over $\mathbf{k}$ is over all points of the quasicontinuum within the first Brillouin zone of the given lattice permitted by periodic boundary conditions. The first term in (3.7) is a non-local part of the chemical free energy and corresponds to the enthalpy of a system. The second term is a local part of the free energy and corresponds to the entropy contribution to the free energy. Using Eq. (3.7) for the free energy the function $b(\mathbf{k})$ can be written as:

$$b(\mathbf{k}) = V(\mathbf{k}) + k_B T\frac{1}{p(\mathbf{r})(1 - p(\mathbf{r}))}. \tag{3.8}$$

The second term in this expression, by definition of the atomic density function that varies from 0 to 1, is already positive and the instability condition ($b(\mathbf{k}) \leq 0$) becomes $|V(\mathbf{k})| \geq k_B T\frac{1}{p(\mathbf{r})(1-p(\mathbf{r}))}$. At some temperature T_c this condition becomes valid and the homogeneous state looses its stability to the periodic fluctuations of concentration with the wave vector $\mathbf{k} = \mathbf{k}_0$ corresponding to a minimum of the function $V(\mathbf{k})$.

3.2. Microscopic kinetic equation

To model the temporal evolution of the atomic density function $p(\mathbf{r}, t)$ and to prototype the kinetics of instability phenomena the microscopic diffusion equation first propose by Khachaturyan[26] can be used:

$$\frac{dp(\mathbf{r}, t)}{dt} = \sum_{\mathbf{r}'} L(\mathbf{r}, \mathbf{r}')\frac{\delta F}{\delta p(\mathbf{r}', t))}, \tag{3.9}$$

where $L(\mathbf{r}, \mathbf{r}')$ is the effective rate of elementary diffusion jump between the nearest lattice sites $\mathbf{r}$ and $\mathbf{r}'$, $F\{p(\mathbf{r})\}$ is functional of the atomic density function, $p(\mathbf{r}, t)$. Summation in Eq. (3.9) is carried out over all N lattice sites of the crystal. In fact, the kinetic matrix, $L(\mathbf{r}, \mathbf{r}')$, depends on the atomic density field created by all atoms at point $\mathbf{r}$,[27,28] but at the growth stage this dependence is reduced to a renormalisation of the effective values of $L(\mathbf{r}, \mathbf{r}')$. The kinetic equation (3.9) approximates the evolution rate by the first non-vanishing term of its expansion with respect to the thermodynamic driving force (small driving force). Eq. (3.9) is significantly non-linear with respect to the atomic density function, $p(\mathbf{r})$, although it is linear with respect to the driving force.

To illustrate the different morphologies of microstructure that can be obtained solving Eq. (3.9) we will consider two cases. The first one is the effect of a long-range Coulomb interaction on the dynamics of the phase separation and ordering in oxides. The second example is the microstructure evolution in Ni-based alloys with elastic interaction.

Simultaneous ordering and phase separation is a very common and well documented phenomenon in metallic alloys, ceramics, minerals and even semiconductors. It occurs during the decomposition phenomenon of a homogeneous disordered phase into a two phase mixture of ordered and disordered phases. In Ref. [29] this phenomenon has been studied in oxides compound. A model 2D system was a ceramic compound undergoing a diffusional phase transformation involving ions of the same sign, cations, whereas the diffusion of ions of the opposite sign, anions, occupying different sublattices was frozen. The interatomic interaction contained two contributions: a finite-range part that favours a decomposition of a homogeneous disordered state to a mixture of ordered and disordered phases and a long-range screened Coulomb interaction between atoms of the same type. To demonstrate the dynamics of the formation of a mesoscopic state in this system, a two-dimensional model on a square lattice has been employed. The Fourier transform of the finite-range interaction was given by

$$V_{fi}(\mathbf{k}) = 2W_1 \left[\cos 2\pi h + \cos 2\pi l\right] + 4W_2 \cos 2\pi h \cos 2\pi l + $$
$$2W_3 \left[\cos 4\pi h + \cos 4\pi l\right], \qquad (3.10)$$

where h and l are related to the reciprocal lattice vector $\mathbf{k}$ by $\mathbf{k} = \frac{2\pi}{a_0} [h, l]$ where a_0 is the lattice parameter of the square lattice. It can be easily seen that the minimum of $V_{fi}(\mathbf{k})$ falls at $\mathbf{k}_0 = \frac{2\pi}{a_0} \left[\frac{1}{2}, \frac{1}{2}\right]$. It means that, at low temperature, the fluctuations of the concentration with this wave vectors should grow. The finite range interaction was also completed by the long-range repulsive screened Coulomb interaction:

$$W_{Coul}(r) = \frac{A}{r} e^{-r/r_D}, \tag{3.11}$$

where r_D is the screening radius, r is the distance and A is a parameter that measures the strength of the interaction. The long-wave asymptote of the Fourier transform of Eq. (3.11) is:

$$V_{Coul}(\mathbf{k}) \approx \frac{4\pi A}{v_0(k^2 + k_D^2)}, \tag{3.12}$$

where $k_D = \frac{2\pi}{r_D}$, and v_0 is the atomic volume. Then, the numerical solution of the microscopic kinetics equation (3.9) with the free energy defined by Eq. (3.7) has been obtained. An example of the structural transformation sequence from disordered phase to a mixture of two coexisting phases is shown in Fig. 6. This figure was obtained by an isothermal aging of the disordered phase of concentration 0.175 at a reduced temperature $T^* = 0.0426$ (the reduced temperature is defined as $T^* = \frac{k_B T}{|V(k_0)_{fi}|}$). In Fig. 6 the grey level represents the different magnitudes of the absolute value of $c\eta$, where c is the local concetration and η is the local long-range order parameter. In this case the local long-range order parameter characterises the degree of order in ordered phase at position $\mathbf{r}$ and varies from 0 (in disordered state) to 1 (in fully ordered state). The atomic density function can be expressed through these two variables as $p(\mathbf{r}) = c(\mathbf{r}) + c(\mathbf{r})\eta(\mathbf{r})$. The reduced time is $t^* = t/\tau$, where τ is the typical time for an elementary diffusion event. It is shown in Fig. 6 that the first stage of this kinetics is a non-stoichiometric congruent ordered phase with antiphase domains following decomposition to the two-phase state. However, the more important and very interesting results of these simulation are dramatic changes in the decomposition dynamics. In systems with only finite-range interactions, the resulting two-phase mixture will continuously coarsen reducing its interfacial energy. In this work it is shown that the long-range isotropic Coulomb interaction stops the coarsening after the ordered particle reach a certain size. Fig. 6 (d)-(f) shows that when all particles achieve the same size, the regular triangular (honeycomb) pattern appears. It was also demonstrated that, by increasing the strength of the Coulomb interaction, the size and the distance between ordered particles decreases, but the morphology of the pattern does not change. This very important result shows that in solid-solid transformation the isotropic long-range interaction produces a regular pattern with the same symmetry as that in the convection or in the reaction-diffusion phenomena.

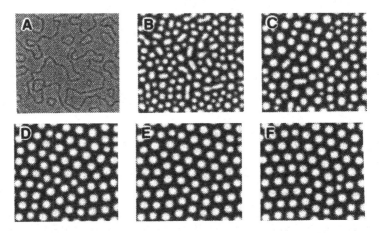

Figure 6. Morphological evolution obtained by numerical solution of Eq. (3.9) with interatomic potential defined by Eqs. (3.12) and (3.10).[29] The input parameters in this simulation are: A=0.25eV, c=0.175, T^*=0.0426, W_1=1eV, W_2=-0.8eV and W_1=-0.55eV. In all figures bright regions represent the ordered domains and dark regions correspond to disordered phase domains at (a) $t^* = 2.5$, (b) $t^* = 10$, (c) $t^* = 100$, (d) $t^* = 500$, (e) t^*=1000 and (f) $t^* = 2000$.

Now let us show what happens in a system during order-disorder phase transformation if a long-range interaction is an anisotropic one. One of the well known examples of this kind of interaction is an elastic interaction. During the past ten years, the effect of elastic strain energy on coherent precipitate morphology and its temporal evolution has been a subject of extensive experimental and theoretical studies (for reviews see Refs. [30, 31]).

In the next example we consider the microstructural evolution in a Ni-V system. In Ni-V systems, a low symmetry ordered phase (tetragonal Ni_3V phase with the DO_{22} ordered structure) coexists with the high symmetry face cubic centered (fcc) disordered parent A1 phase. The crystal lattice of the DO_{22} structure is presented in Fig. 7. In Ni-V systems, the formation of the DO_{22} tetragonal phase produces an anisotropic strain field due to the contraction and expansion of the x,y and z axes against the cubic matrix. This generates the three [100], [010] and [001] orientation variants of the tetragonal phase, all of which are energetically equivalent and thus have the same probability to form during aging from the disordered cubic phase. The resulting microstructure exhibits a fine brick-like "multi-variant structure" (MVS) and the variants are found to be perpendicular twins related to each other across $\{110\}_\gamma$ plane. Many attempts have been made by a few

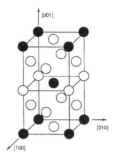

Figure 7.   Unit cell of DO_{22} ordered structure.

authors to model a microstructure with coexisting tetragonal and cubic phases. For example, Y. Le Bouar et al.[32,33] investigated the effect of elastic energy on the morphology of a chessboard-like microstructure in Co-Pt and $(CuAu)_{1-x}Pt_x$. It was shown that the parent cubic phase ($L1_2$ for Co-Pt and disordered fcc for $(CuAu)_{1-x}Pt_x$ coexists with the three different orientation variants of the tetragonal $L1_0$ ordered phase. In this case the final microstructure stays multi-variant and strongly depends on the misfits between two coexisting phases.

In Ni-V alloys, experiments revealed that long-time aging leads to a monovariant microstructure.[34–36] It was shown that during coarsening, the major variant in initial MVS grows at the expense of minor variants and disordered A1 phase.

To investigate the morphological evolution in the Ni-V system, we employed a computer simulation model based on the microscopic kinetics equations Eq. (3.9). In the case of cubic-to-tetragonal transformation the free energy of the system contains two terms:

$$F = F_{chem} + E_{elas},   (3.13)$$

where the first term F_{chem} corresponds to the chemical energy that defines the basic thermodynamic properties of the system and is described by Eq. (3.7) and E_{elas} is the total elastic strain energy.

A thermodynamic model is developed using the static concentration wave (SCW) formalism applied to the DO_{22} structure. This structure is generated by the wave vectors: $\mathbf{k}_1 = (100)$ and $\mathbf{k}_2 = (1/200)$. Using the SCW formalism, the probability distribution function $p(\mathbf{r}, t)$ for the DO_{22} structure can be written as:

$$p(\mathbf{r}(x, y, z)) = c + \frac{1}{4}\eta_1 e^{2i\pi x} + \frac{1}{2}\eta_2 \cos\left(2\pi(\frac{x}{2} + y)\right),   (3.14)$$

where c is the nominal Vanadium concentration of the alloy, η_1 and η_2 are the order parameters, and vary from 0 (disordered state) to 1 (fully ordered state). Substitution of Eq. (3.14) into (3.7) gives the expression for the chemical free energy for a DO_{22} structure:

$$F_{DO_{22}} = \frac{1}{2}Nc^2V(\mathbf{k}=0) + \frac{1}{2}N\left(\frac{\eta_1}{4}\right)^2 V(\mathbf{k}_1) +$$

$$N\left(\frac{\eta_2}{4}\right)^2 V(\mathbf{k}_2) - TS_{DO_{22}}, \tag{3.15}$$

where the entropy S is:

$$S_{DO_{22}} = -Nk_B\left\{\frac{1}{2}\left[(c-\frac{\eta_1}{4})\ln(c-\frac{\eta_1}{4}) + (1-c+\frac{\eta_1}{4})\ln(1-c+\frac{\eta_1}{4})\right] +\right.$$

$$\frac{1}{4}\left[(c+\frac{\eta_1}{4}+\frac{\eta_2}{2})\ln(c+\frac{\eta_1}{4}+\frac{\eta_2}{2}) + (c-\frac{\eta_1}{4}-\frac{\eta_2}{2})\ln(c-\frac{\eta_1}{4}-\right.$$

$$\left.\left.\frac{\eta_2}{2}) + (1-c-\frac{\eta_1}{4}+\frac{\eta_2}{2})\ln(1-c-\frac{\eta_1}{4}+\frac{\eta_2}{2})\right]\right\}. \tag{3.16}$$

Here, N is the total number of sites and $V(\mathbf{k})$ is the Fourier transform of exchange energies defined previously. For a fcc lattice and for the $\mathbf{k}$ vectors that generate the DO_{22} ordered structure, we can write:

$$V(\mathbf{k}=\vec{0}) = 12w_1 + 6w_2 + 24w_3 + 12w_4,$$

$$V(\mathbf{k}_1) = -4w_1 + 6w_2 - 8w_3 + 12w_4,$$

$$V(\mathbf{k}_2) = -4w_1 + 2w_2 + 8w_3 - 4w_4, \tag{3.17}$$

where ω_1, ω_2, ω_3 and ω_4 are the first-, second-, third- and forth-nearest neighbour effective exchange interaction energies.

The elastic energy term was calculated using the elastic theory of multiphase coherent solids proposed in Ref. [24]. In the homogeneous modulus approximation, the elastic strain energy is:

$$E_{elas} = \frac{1}{2}\sum_{p,q}\int{}'\frac{d^3k}{(2\pi)^3}B_{pq}(\mathbf{n})\left\{\eta_p^2(\mathbf{r},t)\right\}_k\left\{\eta_q^2(\mathbf{r},t)\right\}_k^*, \tag{3.18}$$

where function $B_{pq}(\mathbf{n})$ describes the elastic properties and the transformation crystallography whereas the domain morphology and distribution of precipitates (inclusions of the DO_{22} ordered phase) are described by $\left\{\eta_p^2(\mathbf{r})\right\}$. $\mathbf{n} = \frac{\mathbf{k}}{k}$ is a unit vector in the $\mathbf{k}$ direction. Asterisk indicates the complex conjugate function the prime in Eq. (3.18) indicates that $\mathbf{k}=0$ is excluded from the integration. According to Ref. [37], in order to describe the cubic-tetragonal transformation with approximation of isotropic elasticity and strain expansion, the function can be written in 2D in the next form:

$$B_{xx}(\mathbf{n}) = \frac{2\mu\delta_{33}^2}{1-\upsilon}\left(\left[(1-d)n_1^2 - (1+\upsilon d)\right]^2 + (1-\upsilon^2)d^2\right),$$

$$B_{yy}(\mathbf{n}) = \frac{2\mu\delta_{33}^2}{1-\upsilon}\left(\left[(1-d)n_2^2 - (1+\upsilon d)\right]^2 + (1-\upsilon^2)d^2\right),$$

$$B_{xy}(\mathbf{n}) = \frac{2\mu\delta_{33}^2}{1-\upsilon}\left((1-d)^2 n_1^2 n_2^2 + (1+\upsilon)d(1+d)\right), \tag{3.19}$$

where $d = \delta_{11}/\delta_{33}$, $\delta_{11} = \dfrac{a-a_0}{a_0}$ and $\delta_{33} = \dfrac{c/2-a_0}{a_0}$ (a_0, a and c are the crystal lattice parameters of the disordered fcc phase and the ordered DO_{22} phase, respectively).

Local order parameters η_p in Eq. (3.18) can be defined as follows. As we remarked previously, each orientation variant of the DO_{22} unit cell is generated by two vectors. The x-variant is generated by $\mathbf{k}_{1x}(100)$ and $\mathbf{k}_{2x}(\frac{1}{2}00)$ vectors, and y-variant by $\mathbf{k}_{1y}(010)$ and $\mathbf{k}_{2y}(1\frac{1}{2}0)$ vectors. In our simulation we calculate these two local order parameters using a small volume (5×5 sites) around a given atom. In this case the order parameters can be defined as:

$$\eta_{1x}(\mathbf{r},t) = e^{-i\mathbf{k}_{1x}\mathbf{r}}\left(D_{1x} \otimes p(\mathbf{r},t)\right), \tag{3.20}$$

$$\eta_{1y}(\mathbf{r},t) = e^{-i\mathbf{k}_{1y}\mathbf{r}}\left(D_{1y} \otimes p(\mathbf{r},t)\right),$$

with

$$D_{1x} = \begin{pmatrix} 1/4 & -1/2 & -1/2 & -1/2 & 1/4 \\ 1/2 & -1 & 1 & -1 & 1/2 \\ 1/2 & -1 & 1 & -1 & 1/2 \\ 1/2 & -1 & 1 & -1 & 1/2 \\ 1/4 & -1/2 & 1/2 & -1/2 & 1/4 \end{pmatrix} \tag{3.21}$$

and

$$D_{1y} = \begin{pmatrix} 1/4 & 1/2 & 1/2 & 1/2 & 1/4 \\ -1/2 & -1 & -1 & -1 & -1/2 \\ 1/2 & 1 & 1 & 1 & 1/2 \\ -1/2 & -1 & -1 & -1 & 1/2 \\ 1/4 & 1/2 & 1/2 & 1/2 & 1/4 \end{pmatrix}. \tag{3.22}$$

Due to the periodic boundary conditions applied to the simulation box, the
Eq. (3.18) can be rewritten in the discrete form as follows:

$$E_{elas} = \frac{v_0}{2N} \sum_{p,q} \sum_{\mathbf{k} \in B_1} {}' B_{pq}(\mathbf{n}) \left\{ \eta_{1p}^2(\mathbf{r}, t) \right\}_{\mathbf{k}} \left\{ \eta_{1q}^2(\mathbf{r}, t) \right\}_{\mathbf{k}}^*, \qquad (3.23)$$

where v_0 is the fcc unit cell volume, N is the number of sites in simulation
box and the summation is carried out in the first Brillouin zone B_1. With
definitions by Eq. (3.20)–(3.22) the Eq. (3.23) explicitly gives the elastic energy using the probability distribution function $p(\mathbf{r}, t)$. In this form
the elastic energy can be directly integrated in the microscopic diffusion
equation (3.9).

The kinetic equation (3.9) has been numerically solved in reciprocal
space using the explicit forward Euler technique. Two-dimensional simulation was performed with a 768×768 sites box, and corresponds to 135×135
nm. Periodic boundary conditions were applied. Nominal Vanadium concentration was fixed at 20%. The system was started from a homogeneous
disordered state with randomly distributed 1500 small DO_{22} precipitates
(x- and y-variants). According to the experimental results presented in
Ref. [38] we have used the following set of elastic parameters: $T = 800C$,
$\mu = 367$ meV/Å^3, $\nu = 1/3$, $\delta_{11} - 0.0053$ and $\delta_{33} = 0.0106$.

The exchange energies were fitted to the experimental phase diagram.[39]
The free energies of the DO_{22} ordered phase as functions of concentration
at a given temperature are obtained by minimisation of the chemical free
energy, described by the Eq. (3.7), with respect to the order parameters.
Finally, the equilibrium concetrations of ordered and disordered phases were
determined numerically by the common tangent construction. A reasonably
good fit was obtained using the following chemical interaction parameters
with a third-neighbour interaction model: $\omega_1 = 93.33$ meV; $\omega_2 = -27.92$
meV; $\omega_3 = -25.10$ meV.

We suppose that diffusion takes place only between nearest-neighbour
lattice sites, so that $L(\mathbf{r}, \mathbf{r}')$ is equal to constant L_1 if sites $\mathbf{r}$ and $\mathbf{r}'$ are
first-neighbour and 0 for further neighbour order. At the initial step of
simulation, the small precipitates with the DO_{22} structure were artificially
introduced in the simulation box. This implies that information about the
early stages of nucleation, like the incubation time, is not available from
such simulations. This is also why microstructural evolution will be given
as a function of the reduced time ($t^* = L_1 t$). It is worth noting that
this limitation does not have a great importance in the framework of this
research. The aim of simulations was not to study a nucleation regime

but instead the coarsening stage. Particular attention will be paid to the influence of elastic interaction on the coarsening kinetics, and more precisely on the relative stability of orientation variants of Ni_3V precipitates.

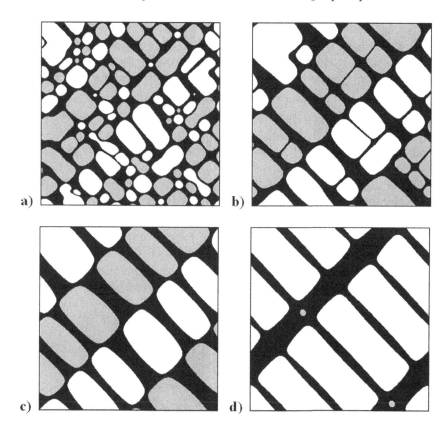

Figure 8. Morphological evolution of DO_{22} precipitates from 2D simulation in a Ni-20%atV at different reduced time : (a) $t^*=4$; (b) $t^*=30$; (c) $t^*=120$ and (d) $t^*=750$. x- and y-variant are represented in different shadow of gray. Disordered fcc matrix is in black. Computational domain size is 135×135 nm.

The microstructural evolution in a Ni-20% at V alloy is presented in Fig. 8. The x- and y-variant are represented in different shadow of gray. The disordered fcc matrix is in black. The initial stage microstructure (Fig. 8 (a): $t^*= 4$) consists of rectangular ordered domains with characteristic alignment along the $\langle 110 \rangle$ directions. x- and y-variants are still mixed together but monovariant $\langle 110 \rangle$ aligned bands of precipitates start to be formed. At $t^*= 30$ (Fig. 8 (b)), the microstructure is composed of alternated monovariant bands aligned along the $\langle 110 \rangle$ direction. At this stage we also observe the independent coarsening in each monovariant bands. At $t^* = 120$ (Fig. 8 (c)), local coarsening is finished and DO_{22} precipitates form chessboard-like structure. The next stage of kinetics involves a non-local coarsening phenomenon. At this kinetics step all of the rows of y-variant precipitates disappear and the structure becomes mono-variant (Fig. 8 (d)). Our simulation results reproduced quite well the experimental microstructure evolution in Ni-V alloy aging from 1 h to 10 h at 800°C (see Fig. 9). Fig. 9 gives a comparison of simulated and experimental chessboard-like structures. We can see that the microstructures are similar as well as the size of precipitates (50 nm).

In order to better understand the influence of the misfit and more precisely the ratio between two misfits $d = \delta_{11}/\delta_{33}$, the simulation with different values of d have been performed. In Fig. 10 the final microstructures obtained for different values of d are presented. We can see that a small variation of misfit ratio drastically changes the morphology of the microstructure. For $d = -0.5$ the equilibrium microstructure corresponds to the mono-variant structure. At $d = -0.35$ the microstructure is described by the chessboard-like morphology. In the system where the tetragonal deformation is very small or d is very low the microstructure can be described by 2D

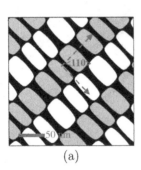   

                    (a)   (b)

Figure 9.   Comparison of precipitate morphologies obtained by simulation (a) and experiment (b): (a) microstructure at $t^* = 150$ in Ni-V, (b) dark field image showing the two DO_{22} variants in Ni-19%atV aged 3 h at 800°C.[40] x- and y-variant are represented in different shadow of gray. Disordered fcc matrix is in black.

orthorhombic particles aligned along the $\langle 110 \rangle$ direction. All of these microstructures can be found in the real systems. In Fig. 11 the experimental images obtained by Transmission Electron Microscopy (TEM) are shown. By comparing these two figures, we can conclude that the proposed model gives a very good description of elastic properties of the Ni-V-X ternary systems and reproduces quite well the morphology of the microstructure with anisotropic elastic interaction.

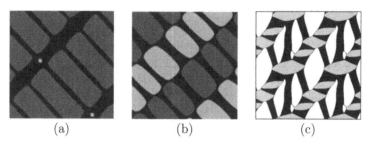

(a)   (b)   (c)

Figure 10. Simulated mictrostructure in Ni-V alloy for different ratio of misfit: (a) $d = -0.5$, (b) $d = -0.35$, (c) $d = 0$.

Analysing these modelling, we can conclude that the long-range anisotropic elastic interaction produces very rich range of morphology of microstructure in alloys. The ordered particles with the DO_{22} structure align to elastically soft directions.

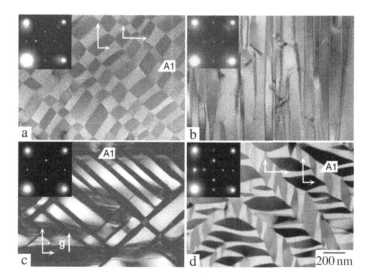

Figure 11. Dark field image showing the two DO_{22} variants microstructure in the ternary alloys aged 864ks at 1073 K:[34] (a) and (b) Ni-20%atV-10%atCo, (c) Ni-19%atV-4%atFe, (d) Ni-15%atV-5%atNb.

3.3. Modelling of solid-solid phase transformation at mesoscopic scale

In most cases, the characteristic length of the microstructure in materials is an order of micrometer and to describe its evolution the mesoscopic scale modelling is required. During the last decades, the phase field model has been extensively used for microstructure simulation studies because of its ability to reproduce complicated microstructures without any a priori assumptions at mesoscopic scale.[42-46] In this approach, each phase or domain in a microstructure is characterised by a set of field variables (compositions and/or order parameters). The basic idea in the phase field model is to introduce an order parameter or phase field that varies continuously over thin interfacial layers and is mostly uniform in the bulk phases. Perhaps the best-known example of this type of model is the Cahn-Hilliard equation[47,48] used for modelling phase separation in a binary mixture quenched into the unstable region. In a general case, the temporal evolution of the field variables is given by the Cahn-Hilliard and Langevin equations:

$$\frac{\partial c(\mathbf{r}, t))}{\partial t} = \bigtriangledown M \bigtriangledown \left(\frac{\partial F}{\partial c(\mathbf{r}, t)} \right) + \xi, \tag{3.24a}$$

$$\frac{\partial \eta_i(\mathbf{r}, t))}{\partial t} = -L \frac{\partial F}{\partial \eta_i(\mathbf{r}, t)} + \zeta, \tag{3.24b}$$

where $\bigtriangledown$ is the differential operator, L and M are kinetic coefficients, F is a total non-equilibrium free energy functional, c is a local concentration, η_i ($i = 1, 2, 3$) represents the components of order parameter, ξ and ζ are the Langevin noise terms describing the compositional and structural thermal fluctuations, respectively.

The total free energy F of the system consists of the incoherent bulk free energy, the interfacial energy (gradient terms) and the elastic energy E_{elas}. Assuming an isotropic interfacial energy, the chemical energy F_{chem} in the diffuse-interface description is:[24]

$$F_{chem} = \int_V \left(f(c, \eta_j) + \frac{\beta}{2} (\bigtriangledown c)^2 + \frac{\alpha}{2} \sum_{j=1}^n (\bigtriangledown \eta_j)^2 \right) dV, \tag{3.25}$$

where α and β are gradient energy coefficients, f is the local free energy density that can be presented in Landau polynomial form and in the case of binary systems can be expressed as:

$$f(c, \eta(\mathbf{r})) = A_1 + \frac{A_2}{2} \eta^2(\mathbf{r}) + \frac{A_3}{3} \eta^3(\mathbf{r}) + \frac{A_4}{4} \eta^4(\mathbf{r}), \tag{3.26}$$

where $A_i \equiv A_i(c, T)$ are the expansion coefficients that determine the thermodynamic properties of system and depend on the temperature and concentration.

Let us consider the simplest case when the phase separation is related to the concentration variations. To investigate the stability of a system with respect to the variations of the concentration, the variation of the free energy with respect to this fluctuation should be evaluated. Using Eq. (3.25) the variation of the free energy is:

$$\Delta F_{chem} = \frac{1}{2} \int_V \left[\frac{d^2 f(c)}{dc^2} \bigg|_{c=\bar{c}} \delta c^2(\mathbf{r}) + \frac{1}{2} \beta(\bar{c}) \left(\nabla \delta c(\mathbf{r}) \right)^2 \right] dV, \qquad (3.27)$$

where $\bar{c}$ is the mean concentration of atoms of a given sort. The Fourier transform of this expression gives:

$$\Delta F_{chem} = \frac{1}{2N} \sum_{\mathbf{k}} b(\mathbf{k}) \left| c(\mathbf{k}) \right|^2, \qquad (3.28)$$

with

$$\delta c(\mathbf{r}) = c(\mathbf{r}) - \bar{c} = \frac{1}{N} \sum_{\mathbf{k}} c(\mathbf{k}) e^{-i\mathbf{k}\mathbf{r}}$$

and

$$b(k) = v \left[\frac{d^2 f(c)}{dc^2} \bigg|_{c=\bar{c}} + \frac{1}{2} \beta(\bar{c}) k^2 \right]. \qquad (3.29)$$

Here v is a volume of the unit cell. As follows from Eq. (2.7) the system becomes unstable with respect to the fluctuation $\delta c(\mathbf{k})$ if $\frac{d^2 f(c)}{dc^2} \big|_{c=\bar{c}} < 0$ and $\left| \frac{d^2 f(c)}{dc^2} \big|_{c=\bar{c}} \right| \geqslant \beta(\bar{c}) k^2$, the second term in Eq. (3.28) being already positive. At a temperature higher than the temperature of phase transition T_c from a disordered homogeneous state to an ordered (periodic) one, the function $b(\mathbf{k})$ is positive for all values of $\mathbf{k}$ and the system is stable with respect to any fluctuations. Then, at $T = T_c$ $b(\mathbf{k}_0) = 0$ and the system becomes unstable to the periodic arrangement of atoms with periodicity $a = 2\pi/\mathbf{k}_0$. With a decreasing temperature the system becomes unstable with respect to some number of wave vectors $\mathbf{k}$. However the fluctuations with $\mathbf{k}=\mathbf{k}_0$ grow faster. In reality there are two different type of instability that can be produced. If the minima of $b(\mathbf{k}_0)$ is reached at $\mathbf{k}_0=0$ it corresponds to the spinodal instability with a miscibility gap. In this case the decomposition reaction pushes to a separation of two different species of atoms at maximum possible distance. If the minimum of $b(\mathbf{k}_0)$ corresponds to a certain value of $\mathbf{k} = \mathbf{k}_0$ then a homogeneous state becomes unstable to ordering or periodic distribution of atoms. The characteristic function $b(\mathbf{k})$ thus contains the information sufficient to predict whether the phase transformation is an ordering or decomposition. From this consideration we can see that the function $b(\mathbf{k})$ is a continuum version of the function $b(\mathbf{k})$ introduced in the microscopic model. The gradient term in the phase field model plays the same role as the interatomic interaction term in the microscopic approach. To link these two approaches we can expand the Fourier transform of the kinetic coefficients $L(\mathbf{k})$ into the Taylor series with respect to $\mathbf{k}$ at $\mathbf{k}=0$:

$$\tilde{L}(\mathbf{k}) = \tilde{L}(0) + \sum_i \frac{\partial \tilde{L}(\mathbf{k})}{\partial k_i}\bigg|_{\mathbf{k}=0} k_i + \frac{1}{2} \sum_{i,j} \frac{\partial^2 \tilde{L}(\mathbf{k})}{\partial k_i \partial k_j}\bigg|_{\mathbf{k}=0} k_i k_j + \dots . \quad (3.30)$$

Using the conservation law

$$\int_V L(\mathbf{r} - \mathbf{r}')d^3\mathbf{r} = \tilde{L}(0) = 0$$

and isotropy of the space $(L(\mathbf{k}) = L(-\mathbf{k}))$ then only the even powers of k in Eq. (3.30) should be considered. Then, taking only the first non-vanishing term in the Eq. (3.29) the Fourier transform of the kinetic coefficient is:

$$\tilde{L}(\mathbf{k}) = -\sum_{ij} M_{ij} k_i k_j, \quad (3.31)$$

where

$$M_{ij} = -\frac{1}{2} \frac{\partial \tilde{L}(\mathbf{k})}{\partial k_i \partial k_j}\bigg|_{\mathbf{k}=0} .$$

Here M is a usual definition of the atomic mobility. Replacing $L(\mathbf{k})$ in Eq. (3.9) gives:

$$\frac{\partial \tilde{p}(\mathbf{k}, t)}{\partial t} = -\sum_{ij} M_{ij} k_i k_j \left[\frac{\delta F}{\delta p(\mathbf{r}', t)} \right]_{\mathbf{k}} . \quad (3.32)$$

The inverse Fourier transform of this differential equation is:

$$\frac{\partial p(\mathbf{r}, t)}{\partial t} = -M \nabla^2 \frac{\delta F}{\delta p(\mathbf{r}, t)} . \quad (3.33)$$

Now, in this equation the variable $p(\mathbf{r}, t)$ is the continuous variable. The coarse grain of this variable $p(r)$ on some small volume (it can be, for example, the volume of one crystal lattice cell) will transform this equation to the Cahn-Hillard equation (3.24a).

In the phase field model the elastic interaction is usually described by Khachaturyan's model introduced previously (see Eq. (3.18)). In the continuous case it can be written in the next form:

$$E_{elas} = -\frac{1}{2} \int_V B(\mathbf{n}) \, |c(\mathbf{k})|^2 \, \frac{d^3k}{(2\pi)^3}, \tag{3.34}$$

where the integral is over the reciprocal Fourier space, $c(\mathbf{k})$ is the Fourier transform of composition field $c(\mathbf{r})$. The function $B(\mathbf{n})$ contains all information on the elastic properties of the systems and depends on the elastic constants and the coefficient of lattice expansion caused by the changes in the composition. In a more simple case where precipitates and matrix have the same cubic symmetry the elastic properties of a system can be described by only three elastic constants: C_{11}, C_{12} and C_{44}. In this case the decomposition of a cubic phase with the crystal lattice parameter a_0 into a two-phase mixture of cubic phases is characterised by a simplest crystallography because the crystal lattice rearrangement is described by the stress-free transformation strain:

$$\varepsilon_{ij}^0 = \varepsilon_0 \delta_{ij}, \tag{3.35}$$

where δ_{ij} is the Kronecker delta symbol and the lattice expansion coefficient with respect to concentration ε_0 can be calculated using the lattice misfit δ_0:

$$\delta_0 = \frac{a_p - a_0}{a_0} . \tag{3.36}$$

Here a_p is the crystal parameter of the cubic phase precipitates. If the Vegard law holds, the parameter ε_0 can be written in terms of the concentration coefficient of the crystal lattice expansion. For a binary alloy it is given by:

$$\varepsilon_0 = \frac{\delta_0}{c_p - c_0}, \tag{3.37}$$

where c_p and c_0 is the concentration of solute atoms in the precipitate and matrix, respectively. The stress free tensor ε_{ij}^0 describes the macroscopic shape deformation of the parent phase caused by crystal lattice rearrangements associated with the phase transformation in the stress free state.

Usually, to solve Eqs. (3.24a) and (3.24b) numerically, we use their dimensionless form which can be obtained by dividing both sides of the equations by $L \, |\triangle f|$. If we consider the length l as the space increment of the simulation grid, all length vectors are measured in unit l, i.e. $\mathbf{r} = l\mathbf{r}^*$

where r^* is the reduced coordinate vector. The reduced time can be defined as $t^* = L |\triangle f| t$. All the other parameters can be expressed in their reduced form: $f^* = f / |\triangle f|$, $\nabla^* = l\nabla$ and $(\nabla^*)^2 = l^2\nabla^2$ that are the differential operator and Laplacian in reduced coordinates, the two gradient energy coefficients $\alpha^* = \frac{\alpha}{|\triangle f|l^2}$ and $\beta^* = \frac{\beta}{|\triangle f|l^2}$, the diffusion parameters are $L^* = 1$ and $M^* = \frac{M}{Ll^2}$,the elastic parameters $B(\mathbf{n}) = \mu B^*(\mathbf{n}^*)$ with $\mu = \frac{(c_{11}+2c_{12})^2\varepsilon_0^2}{|\triangle f|c_{11}}$ and $B^*(\mathbf{n}^*) = \frac{c_{11}}{(c_{11}+2c_{12})^2}B(\mathbf{n})$, $\zeta^*(r^*,t^*)$ and $\xi^*(r^*,t^*)$ are the noise functions. To illustrate the phase field method the simulations of the evolution of microstructure in Ni-Al alloys have been performed. For the particular case of conventional Ni-based alloys, the coarsening of γ' (Ni$_3$Al) precipitates in a matrix has been investigated extensively experimentally and by computer simulations.[49–51] According to these works a lattice mismatch between γ' and γ phase which induces the elastic strains in Ni-Al alloys, engenders the morphological changes from a spherical to a cuboidal shape as the particles grow to a larger size. Furthermore, the elastic interactions are infinitely long-range, this causes strong spatial correlations between the precipitates leading to their alignment along certain crystallographic directions. Finally, the coarsening of ordered intermetallic precipitates is further complicated by the fact that they can exist in several types of anti-phase domains related by a lattice translation to the parent disordered phase. To understand separately the influence of the elastic contribution and the presence of anti-phase domains between the ordered γ' particles on the coarsening kinetics, the precipitation of particles from supersaturated, off-stoichiometric γ' matrix has been studied. Following Y. Ma and A.J. Ardell[52] we will call these types of alloys inverse alloys. In this case there is no anti-phase relationship among precipitates; however, the lattice misfit strain is identical with the conventional Ni-based alloys.

Then in inverse Ni-Al alloy the matrix has the $L1_2$ ordered structure which is based on the fcc unit lattice where Al atoms occupy (000) site and Ni atoms are in (1/2,1/2,0), (1/2,0,1/2) and (0,1/2,1/2) positions. To describe this ordered structure three order parameters are required. In other words, the unit cell of the ordered structure contains four sublattices and to describe the distribution of atoms at these sublattices only three order parameters, which will distinguish these sublattices, should be introduced. In the case of the $L1_2$ structure three of four sublattices are occupied by the same type of atoms this way we can consider that $\eta_1 = \eta_2 = \eta_3 = \eta$.

In this case the chemical free energy of this alloy can be written as:

$$f(c,\eta) = \frac{A_1'}{2}(c - c_1)^2 + \frac{A_2'}{2}(c_2 - c)\eta^2 + \frac{A_3}{3}\eta^3 + \frac{A_4}{4}\eta^4. \qquad (3.38)$$

Here c is the solute composition, c_1 and c_2 are constants with values close to equilibrium compositions of the matrix and precipitates, respectively, A_i and A_i' are the constants for the given temperature. In a given example, the coefficients and constants were chosen such that the free energy curve provides a qualitative description of the thermodynamics of Ni-Al system at 1000K, with the constraint that A_3 is non-zero for the order-disorder phase transition to be first order. The free energy curve as a function of aluminum concentration is shown in Fig. 12. This function is obtained by minimising Eq. (3.36) with respect to the order parameter. This curve was obtained using the next set of parameters[50] and expressed in units of $3.35 \cdot 10^7$ J/m^3: $A_1' = 277.78$, $A_2' = 66.67$, $A_3 = -21.21$, $A_4 = 22.14$, $c_1 = 0.1123$ and $c_2 = 0.2211$, $c*$ is the concentration at which the γ' and γ' phases have the same free energy and $|\triangle f|$ is the typical driving force of the order-disorder transformation. In our simulation $|\triangle f| = 0.2$. Simulations were

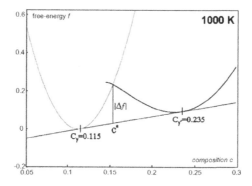

Figure 12. Free energy curves of the disordered γ and ordered γ' phase at T=1000K, expressed in units of $3.35 \cdot 10^7$ J/m^3.

performed by numerically solving the four nonlinear equations (Eq. (3.24a) and Eq. (3.24b)), one for concentration and three equations for the three order parameters, using the semi-implicit Fourier-Spectral method.[53] In 2D computer simulations, 1024×1024 discrete grid points are used and periodic boundary conditions are applied along x and y axes. A uniform time step $\triangle t^*$ was chosen to be 0.1 and a mesh size of $\triangle x=1.0$. The initial state was a homogeneous solution with small composition fluctuations around the average composition. Spherical nuclei with an average radius greater than a nucleation radius were randomly introduced to the computation domain. Since our focus is only on the coarsening kinetics and not the sequence of transformations, the artificially chosen initial configuration does not affect our results. The gradient energy coefficient β (assuming α to be zero) in Eq. (3.25) is chosen to be $4 \cdot 10^{-11}$ J/m, which gives rise to an interfacial energy of 25 mJ/m^2 based on Cahns theory.[47] The coefficient $B(\mathbf{n})$ is estimated as $120 \cdot 10^{10}$ J/m^3, based on the elastic constants of Ni solid solution: C_{11}=209 GPa, C_{12}=149 GPa and C_{44}=96 GPa at 1000K. The parameter ε_0 may be estimated from γ/γ' lattice misfit and it gives ε_0= 0.046.

Two precipitate volume fractions of γ phase, 0.2 and 0.3, were studied with the same kinetic parameters at 1054K to determine the effect of volume fraction on the morphology and coarsening kinetics. The temporal microstructural evolution for both alloys during coarsening is shown in Fig. 13. The coherent disordered γ precipitates are shown in shadow of gray in and the γ' ordered matrix is shown in white. With increasing of aluminum concentration the colour of the matrix goes to black. For both volume fractions, we observe that the anisotropic elastic energy induced cuboidal shape for precipitates and the anisotropic elastic interaction induced alignment along elastically soft [10] and [01] directions even for the precipitates with the small size. This is different from what occurs in conventional alloys. Wang and Khachaturyan[54] were the first to suggest that anti-phase relationships among nearest neighbour γ' precipitates were responsible for their resistance to coalescence. In the case of inverse alloys, it is impossible for disordered precipitates to be anti-phase and the neighboring γ particles coalesce very rapidly.

These simulation results are in good agreement with experimental data reported by Ma and Ardel.[52] In this study, the coalescence in Ni-Al inverse alloys containing 22.25, 22.33 22.40 22.60 and 22.80 %at Al aged at 650°C using experimental transmission electron microscope images was investigated. It corresponds to precipitate volume fraction that varies from 0.7 to 6. Unfortunately, it is impossible to reproduce in our simulation the kinetics in the system with a very small volume fraction of precipitates: this is why we can compare our results with experimental data only qualitatively. However, this qualitative comparison gives an insight on the influence of anti-phase boundaries on coalescence kinetics.

In two-phase solids, the dynamics of coarsening is governed also by the interfacial and elastic energies. This can be shown by examining the

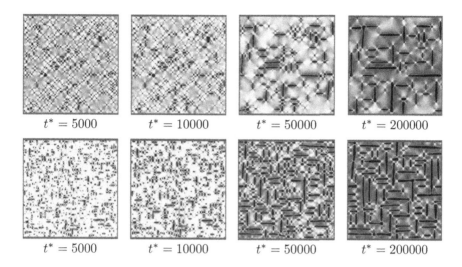

$t^* = 5000$ $\qquad$ $t^* = 10000$ $\qquad$ $t^* = 50000$ $\qquad$ $t^* = 200000$

$t^* = 5000$ $\qquad$ $t^* = 10000$ $\qquad$ $t^* = 50000$ $\qquad$ $t^* = 200000$

Figure 13. Simulated microstructure at different dimensionless time step t^* in Ni-Al alloys with 20% (upper row) and 30% (low row) precipitate equilibrium volume fraction. The intensity of gray colour is proportional to the level of aluminium concentration.

magnitude of dimensionless parameter, which is a measure of the relative importance of elastic and interfacial energies in the system. This parameter for a system with multiple particles was given by P.W. Voorhees[56] and can be written as:

$$\langle L \rangle = \frac{\varepsilon_0^2 \langle l \rangle c_{44}}{\sigma}, \qquad (3.39)$$

where $\langle l \rangle$ is an average characteristic length associated with the particles in the system and σ is the interfacial energy. It was shown in [56] that for conventional Ni-Al alloys there is the bifurcation point at $\langle L \rangle$=5.6 that corresponds to the changing of the particle shape from fourfold to twofold symmetry. It was found also that in the case of multiparticle systems, the elastic interaction between the precipitates perturbs the shape of an individual particle and the cuboidal particles with fourfold symmetry are observed for $\langle L \rangle$ bigger than 5.6 and the elongated forms of particles with twofold symmetry are observed for $\langle \mathbf{L} \rangle$ smaller than 5.6. There results have shown the scattering of values the curvature of forms of precipitates around some mean curve. To understand the influence of elastic interaction between the particles for the non-conventional Ni-Al alloys, the aspect ratio $A = (x - y)/(x + y)$ of particles as function of the reduced equivalent radius R^* has been calculated. For each precipitate, we calculated the values x

and y which correspond to the size of particle along [10] and [01] directions, respectively. In most cases in observed microstructure, the particles have rectangular shapes and the value of A gives a good measurement of particle symmetry. A positive value of the aspect ratio indicates the elongation of a particle to [10] direction, and negative value corresponds to the elongation of a particle to [01] direction. Figs. 14 (a) and 14 (b) show the aspect ratio as a function of R^* for different volume fractions of precipitates. For each figure the results from the different time steps are interposed. It should be pointed out that at small values of radius (between 0 and 10) some number of discreet branches is observed. This effect is an artefact related to the discretisation of computational grid. As a first-order comparison to results from conventional and non-conventional Ni-Al alloys we can conclude that the absence of antiphase boundaries between the precipitates does not change the general behaviour of A as function of R^*. We observed for small values of R^* that the precipitates have cubic and spheroidal forms. This implies that there is a small scattering for A value for the small value of R^*. With an increase of R^* the particles elongate to two elastically soft directions and we observe two branches (positive and negative), Figs. 14 (a) and 14 (b). The large dispersion of the A values for $R^* > 8$ is due to two reasons: first, the coalescence of two elongated perpendicular particles produces the T-type conformation. The aspect ratio of these precipitates gives us a value near 0 which clearly does not reproduce the real configuration shape. These cases are rare but contribute to a scattering of the A values around a mean value of A. Second, as was discussed in [56], in the multiparticles systems the surrounding particles affect individual particle morphology. In real microstructure two particles or more are relatively close or $\langle L \rangle$ is large and the interactions cause the shapes to be distorted significantly. It means that the local configuration of particles determines the shape of an individual particle. With increasing of precipitates volume fraction, the distance between the particles decreases and the influence of neighbouring particles becomes stronger.

Already it could be suggested from our simulation results that the value of the critical equivalent radius, where the majority of the particles change their forms from four to two-fold shapes, does not depend on the precipitate volume fraction and is invariant with time. Following the previous works[55,56] on the morphology particles in large-scale simulations, the value of the critical radius does not effectively depend on time. In this work the critical bifurcation point has been found equal to 5.6.

To relate our value of critical radius with the bifurcation value of

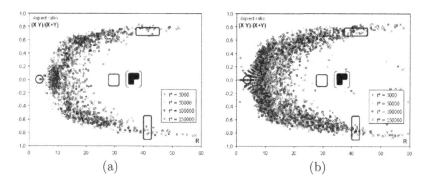

(a)   (b)

Figure 14.   Morphology evolution plot showing the value of the aspect ratio versus reduce radius of each γ' precipitate for four step of the simulation in the two cases: 20% (a) and 30% (b) of γ' phase.

Voorhees parameter L, we first calculated the dimension of the computation grid in real space. Comparing the value of our reduced parameter β^* which is 0.75 and the value of the real parameter β which is 4.10^{-11} J/m^2 we can calculate the value of the scale parameter l which is the increment grid's size of our phase field, considering $|\Delta f| = 0.67 \cdot 10^7$ J/m^3 :

$$l = \sqrt{\frac{\beta}{|\Delta f| \, \beta^*}} = 2.82 \cdot 10^{-9}.$$ (3.40)

Using the real value of the computational grid, the size of the simulation cell is 2.9 nm. Fig. 15 shows the simulation microstructure of the Ni-21.1%Al alloy for $t^* = 10000$ and the dark field TEM image of Ni-22.33%Al.[52] To compare these results, only 25% of the computation box is presented. The size of γ' precipitates obtained from our simulations is in a good agreement with experimental one. If we substitute the value of l in the Eq. (3.39) we can calculate the critical radius:

$$R^* = \frac{\sigma L}{c_{44} \varepsilon_0^2 l} \,.$$ (3.41)

In this calculation, the next values have been used: $\sigma = 25 \cdot 10^{-3} J/m^2$, $L = 5.6$, $\varepsilon_0 = 0.046$ and $c_{44} = 96 \cdot 10^9 J/m^3$. Then bifurcation points correspond to $R^* = 0.248$ and do not really correspond to diagram plotted in Fig. 14. This difference can be caused by our calculation method for precipitates aspect ratio. With this method, small T-shape particles are taken to account in these diagrams but they are assimilated as normal particles, like we have seen with large T-shape particles in the latest stages.

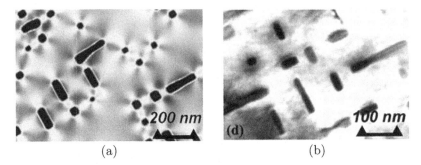

Figure 15. (a): Concentration field of Al in simulated microstructure in Ni-21%Al alloy
$(t^* = 10000)$.[57] The intensity of gray colour is proportional to concentration of the
aluminium atoms. (b): Dark field TEM micrographs of precipitates for an alloy with
22.33% in Al.[52]

We can see that these types of particles are visible for the first stages on the
concentration field. Many small T-shape particles with an equivalent radius
near the critical value could appear in the diagrams and then shift the real
bifurcation curve to the right. In this section we studied the coarsening
kinetics in the inverse Ni-Al alloys, where disordered particles precipitate
in a disordered monovariant $L1_2$ ordered matrix. From these results we can
conclude that when the coherent elastic strain effect is taken into account,
a wide variety of morphological states can be developed at different stages
of a precipitation reaction in alloy. A shape transformation of a single
coherent particle during strain-induced coarsening caused by a competition
between the elastic and interfacial energies was also observed.

4. Discussion and conclusions

The different examples of the pattern formation presented in this chapter
show that the symmetry of the interaction potential plays an important
role in the morphology of pattern. It was deduced that one mode isotropic
potential produces the same type of pattern in completely different types of
systems (chemical reactions, convection, superconductors and microstructure in oxides). In the 2D case we already observe a stripe or honeycomb
structure.

   To understand it, we can establish the relation between the function
$b(\mathbf{k})$, which is responsible for loss of stability of a given system to infinites-
imal fluctuations, and the intensity of the elastic diffuse x-ray or neutron
scattering. In the case of a multicomponent system this function is a $n \times n$

matrix $b_{\alpha\beta}(\mathbf{k})$, where n is the number of components. As we mentioned before, this function is a thermodynamic function and, in the general case, depends on the temperature and pressure. Here we will consider only the system at normal conditions and we will only discuss the influence of temperature on phase stability. The intensity of the elastic diffuse scattering in the kinematic (Born) approximation is:

$$I(\mathbf{k}) = \sum_\alpha \sum_\beta f_\alpha(\mathbf{k}) f_\beta(\mathbf{k}) \langle Q_\alpha(\mathbf{k}) Q_\beta^*(\mathbf{k}) \rangle, \qquad (4.1)$$

where $\langle \ldots \rangle$ is the thermodynamic average, $f_\alpha(\mathbf{k})$ is the atomic scattering factor of an atom of the component α, $\mathbf{k} = \mathbf{k}_1 - \mathbf{k}_2$ is the scattering vector, and $\mathbf{k}_1$ and $\mathbf{k}_2$ are the wave vectors of the scattered and incident waves, respectively. The correlation function $\langle Q_\alpha(\mathbf{k}) Q_\beta^*(\mathbf{k}) \rangle$ can be expressed through the function $b_{\alpha\beta}(\mathbf{k})$ at high temperatures by using the thermodynamic theory of fluctuations:[58]

$$\langle Q_\alpha(\mathbf{k}) Q_\beta^*(\mathbf{k}) \rangle = k_B T b_{\alpha\beta}^{-1}(\mathbf{k}), \qquad (4.2)$$

where $b_{\alpha\beta}^{-1}(\mathbf{k})$ is the matrix inverse to $b_{\alpha\beta}(\mathbf{k})$. Because of the isotropy of the space, this matrix depends on the modulus of k only. Therefore, $b_{\alpha\beta}(\mathbf{k})$ vanishes not at the one point $\mathbf{k} = \mathbf{k}_0$ but at all the points on entire surface of the sphere with the radius $\mathbf{k}_0$. The wave vectors of the sphere of radius $\mathbf{k}_0$ form a star of the dominant waves (see Fig. 16 (a)). Given the spherical degeneration of the instability density modes, the heterogeneous periodic structure will be generated not by a sole dominant wave but a superposition of all dominant waves.

In fact, the function $b_{\alpha\beta}(\mathbf{k})$ equals to the interaction potential shifted by the value of the second derivative of the free energy local part with respect to fluctuations along the temperature axis (see Eq. (3.8)). Then Fig. 16 (a) can be interpreted as a 2D projection of the interaction potential which is, in this case, the one mode isotropic potential. In this figure the negative values of the function $b_{\alpha\beta}(\mathbf{k})$ are shown in shadow of gray.

Let us consider a small undercooling of system below T_c that $b_{\alpha\beta}(\mathbf{k})$ assumes a negative value only within a narrow spherical layer in the k space around the sphere of the radius k_0: the thickness of this instability layer is $\Delta \mathbf{k}$ and $\Delta \mathbf{k}/\mathbf{k}_0 << 1$. The smaller this undercooling, the smaller the ratio $\Delta \mathbf{k}/\mathbf{k}_0$, and the smaller the number of instability waves that has to be eliminated to transform the homogeneous state to the heterogeneous periodic state. On the other hand, the greater the number of the retained instability waves that become the reciprocal lattice vectors of the first coordination shell, the lower the periodic-stabilising harmonic term in the free energy Eq. (3.4).

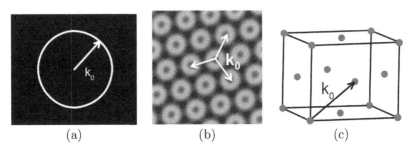

(a)   (b)   (c)

Figure 16.   Stability of the honeycomb and bcc structures. (a) 2D image of the function $b_{\alpha\beta}(\mathbf{k};T)$, (b) unit cell of the honeycomb structure, (c) reciprocal lattice of the bcc structure.

Therefore, in the case of the narrow range of $\Delta\mathbf{k}$ around $\mathbf{k}_0$, the most stable high temperature modification of the periodic phase is the one whose reciprocal lattice has maximum possible number of the nearest neighbour sites. This is the bcc crystal lattice whose reciprocal lattice has 12 nearest neighbour sites (see Fig. 16 (c)). The period of the bcc crystal is related to the vector k_0 as $a = \frac{2\pi}{k_0}\sqrt{2}$. In the case of the 2D system, this is the honeycomb lattice whose reciprocal lattice has six nearest neighbour sites (see Fig. 16 (b)). The period of the honeycomb lattice is expected to be $a = \frac{2\pi}{k_0}$. In the macroscopical systems the interaction between the constituents in most cases can be described by one mode isotropic potential which is a result of the competition between the short-range repulsion and the long-range attraction interactions. This is why at this level most of the patterns have only three types of morphology: stripe, honeycomb or (in 3D) bcc structure. At a mesoscale, where the anisotropic elastic interaction plays an important role, the morphology of the microstructure becomes very rich. At the atomic level, when the interaction between the atoms becomes strongly anisotropic and, in most cases, multi-mode one, the symmetry of the low-temperature phases can be very different from the bcc structure. For example, the function $b(\mathbf{k})$ at a given temperature may have spurious minimum or multiple negative minima associated with the Friedel oscillations in metals. This can explain a large diversity of the crystal structure of metals. In the case of strong undercooling the function $b(\mathbf{k})$ can be negative within a broad range of wave vectors $\mathbf{k}$ and in this case the glass state is formed.

To conclude this chapter we would like to underline again the role of the symmetry of interaction between the constituents of a given system on the symmetry of heterogeneous periodic low temperature state. We hope that our concept will facilitate understanding of many complex patterns.

References

1. M.C. Cross and P.C. Hohenberg, *Pattern formation outside of equilibrium* Rev. Mod. Phys. **65**, 851-1112 (1993).
2. D. Walgraef, *Spatio-Temporal Pattern Formation: With Examples from Physics, Chemistry, and Materials Science (Partially Ordered Systems)* (Springer 1996).
3. J.S. Langer, *Instabilities and pattern formation in crystal growth*, Rev. Mod. Phys. **52**, 1-28 (1980).
4. E. Schäffer, T. Thurn-Albrecht, T.P. Russell, and U. Steiner, *Electrically induced structure formation and pattern transfer*, Nature **403**, 874-877 (2000).
5. H.Nakao and A.S. Mikhailov, *Turing pattern in network-organized activator-inhibitor systems*, Nature Physics **6**, 544-550 (2010).
6. A.H. Reddi, *Role of morphogenetic proteins in skeletal tissue engineering and regeneration*, Nature Biotechnology **16**, 247-252 (1998).
7. E. Karsenti, *Self Organization in cell biology: A brief history*, Nature Reviews: Molecular Cell Biology **9**, 255-262 (2008).
8. J.T. Trujillo Bueno, N. Shchukina, and A. Asensio Ramosa, *Substantial amount of hidden magnetic energy in the quiet Sun*, Nature **430**, 326-329 (2004).
9. A.P. Ingersoll, P.J. Gierasch, D. Banfield, and A.R. Vasadava, *Galileo Imaging Team Moist convection as an energy source for the large-scale motions in Jupiter's atmosphere,* Nature **403**, 630-632 (2000).
10. M.W. Sneddon, J.R. Faedee, and T. Emonet, *Efficient modelling, simulation and coarse-graining of biological complexity with NFsim*, Nature Methods **8**, 177-183 (2011).
11. A.S.G. Curtis, M. Dalby, and N. Gadegaard, *Cell signaling arising from nanotopography: implications for nanomedical devices*, Future Medecine **1**, 67-72 (2006).
12. E. Bodenschatz, J.R. de Bruyn, G. Ahlers, and D.S. Cannell, *Transitions between patterns in thermal convection*, Phys. Rev. Lett. **67**, 3078-3081 (1991).
13. A.M. Turing, *The chemical basis of morphogenesis*, Phil.Trans.of Royal Soc. of London, Series B **237**, 37-72 (1952).
14. JBL Bard, *A model for generating aspects of zebra and other mammalian coat patterns*, Journ. Theor. Biology, **93**, 363-385 (1981).
15. K.R. Mecke and D. Stoyan, *Statistical Physics and Spatial Statistics: The Art Of Analyzing And modelling Spatial Structures And Pattern Formation*(Springer-Verlag Berlin Heidelberg 2000).
16. A. Deutsch and S. Dormann, *Cellular automaton modelling of biological pattern formation* (Birkhäuser Boston 2007).

17. M. Karplus and J.A. McCammon, *Molecular dynamics simulations of biomolecules*, Nature Structural Biology **9**, 646-652 (2002).

18. U. Essmann and H. Trauble, *The direct observation of individual flux lines in type II superconductors*, Phys. Lett. A **24**, 526-527 (1967).

19. H. F. Hess, R. B. Robinson, R. C. Dynes, J. M. Valles, and J. V. Waszczak, *Scanning-Tunneling-Microscope Observation of the Abrikosov Flux Lattice and the Density of States near and inside a Fluxoid*, Phys. Rev. Lett. **62**, 214-219 (1989).

20. J. Murray, *Mathematical Biology* (Springer-Verlag New York Inc. 2008).

21. J. Swift and P. C. Hohenberg, *Hydrodynamic fluctuations at the convective instability*, Phys. Rev. A **15**, 319-328 (1977).

22. L.D. Landau and M. Khalatnikov, *On the anomalous absorption of a sound near to points of phase transition of the second kind*, Dokl. Akad. Nauk SSSR **96**, 469 (1954).

23. L.D. Landau, *Zur Theorie der Phasenumwandlungen II*, Phys. Zurn. Sowjetunion **11**, 26-35 (1937).

24. A.G. Khachaturyan, *Theory of Phase Transformations and Structure of Solid Solutions*, (Nauka Moscow 1974). Reedited A.G. Khachaturyan *Theory of structural transformations in solids*. (New York: Wiley; 1983).

25. P. Weiss, *L'hypothèse du champ moléculaire et la propriété ferromagnétique*, Journ. Phys. Theor. Appl. **6**, 661690 (1907).

26. A. Khachaturyan, *Microscopic theory of diffusion in crystalline solid solutions and the time evolution of the diffuse scattering of X rays and thermal neutrons*, Sov. Phys. Solid State **9**, 2040-2046 (1968) (Engl.Transl. of Fiz. Tverd. Tela).

27. A. G. Khachaturyan, *The problem of symmetry in statistical thermodynamics of substitutional and interstitial ordered solid solutions*, Phys. Stat. Sol. (b) **60**, 9-37 (1973).

28. A. G. Khachaturyan, *Ordering in Substitutional and Interstitial Solid Solutions*, Prog. Mater. Sci. **22**, 1-150 (1978).

29. L. Q. Chen and A. G. Khachaturyan, *Dynamics of simultaneous ordering and phase separation and effect of long-range Coulomb interactions* Phys. Rev. Lett. **70**, 1477-1480 (1993).

30. Y. Wang, L. Q. Chen, and A. G. Khachaturyan, *Kinetics of Strain-Induced Morphological Transformation in Cubic Alloys with a Miscibility Gap*, Acta Metall. Mater. **141**, 279-296 (1993).

31. Y. Z. Wang, L. Q. Chen, and A. G. Khachaturyan, *Shape Evolution of a Coherent Tetragonal Precipitate in Partially-Stabilized Cubic ZrO2 :a Computer-Simulation*, Journ. of the American Ceram. Soc. **76**, 3029-3033 (1993).

32. Y. Le Bouar, A. Loiseau, and A.G. Khachaturyan, *Origin of chessboard-like structures in decomposing alloys. Theoretical model and computer simulation*, Acta Mater. **46**, 2777-2788 (1998).

33. D. Alloyeau, C. Ricolleau, C. Mottet, T. Oikawa, C. Langlois, Y. Le Bouar, N. Braidy, and A. Loiseau, *Size and shape effects on the order–disorder phase transition in CoPt nanoparticles*, Nature Materials **8**, 940-946 (2009).

34. A. Suzuki, H. Kojimi, T. Matsuo, and M. Takeyama, *Alloying effect on stability of multi-variant structure of Ni_3V at elevated temperatures*, Intermetallics **12**, 969-975 (2004).

35. A.Suzuki , M. Takeyama, and T. Matsuo, *A1 → $D0_{22}$ Transformation and Multi-Variant Structure of Ni_3V*, Mat. Res. Symp. Proc. **753**, 363-369 (2003).

36. J.B. Singh , M. Sundararaman, P. Mukhopadhyay, and N. Prabhu, *Evolution of microstructure in the stoichiometric Ni–25at.%V alloy*, Intermetallics **11**, 83-92 (2003).

37. H. Zapolsky, S. Ferry, X. Sauvage, D. Blavette, and L.Q. Chen, *Kinetics of cubic-to-tetragonal transformation in Ni–V–X alloys*, Phil. Mag. **90**, 337-355 (2010).

38. D.K. Na D, J.B. Cohen, and P.W. Voorhees, *Coherent phase equilibrium in the Ni-V system*, Mater. Sci. Eng. **A272**, 10-15 (1999).

39. M.F. Singelton , J.L. Murray, and P.Nash, *Binary Alloy Phase Diagrams*, (eds Massalski TB, Muray JL,Bennet LH, Baker H, Metals Park, Ohio: American Society for Metals, 1986).

40. H. Zapolsky, C. Pareige, D.Blavette, and L.Q. Chen, *Atom probe analyses and numerical calculation of ternary phase diagram in Ni-Al-V system*, CAL-PHAD **21**, 125-134 (2001).

41. L.Q. Chen, *Phase-field models for microstructure evolution*, Ann. Rev. of Mater. Res. **32**, 113-140 (2002).

42. P. Fratzl, O. Penrose, and J. L. Lebowitz, *Modelling of phase separation in alloys with coherent elastic misfit*, Journ. Stat. Phys. **95**, 1429-1503 (1999).

43. Yu. U. Wang, Y. M. Jin, and A. G. Khachaturyan, *Phase field microelasticity theory and modelling of elastically and structurally inhomogeneous solid*, Acta Mater. **92**, 1351-1360 (2002).

44. J. S. Lowengrub, A. Rätz, and A. Voigt *Phase-field modelling of the dynamics of multicomponent vesicles: Spinodal decomposition, coarsening, budding, and fission*, Phys. Rev. E **79**, 031926; 1-13 (2009).

45. Y. U. Wang, Y. M. Jin, A. M. Cuitiño, and A. G. Khachaturyan *Phase field microelasticity theory and modelling of multiple dislocation dynamics*, Journ. Appl. Phys. **78**, 2324-2326 (2001).

46. K. Dayal and K. Bhattacharya, *A real-space non-local phase-field model of ferroelectric domain patterns in complex geometries*, Acta Mater. **55**, 1907-1917 (2007).

47. J. W. Cahn and J. E. Hilliard, *Free energy of a nonuniform system. I. Interfacial free energy*, Journ. Chem. Phys. **28**, 258-267 (1958).

48. J.W. Cahn, *On spinodal decomposition*, Acta Metal. **9**, 795-801 (1961).

49. TM Pollock and S. Tin, *Nickel-Based Superalloys for Advanced Turbine Engines: Chemistry, Microstructure and Properties*, Journ. Prop.Pow. **22**, 361-374 (2006).

50. A. van de Walle, G. Ceder, and U. V. Waghmare, *First-Principles Computation of the Vibrational Entropy of Ordered and Disordered Ni3Al*, Phys. Rev. Lett. **80**, 4911-4914 (1998).

51. V. Vaithyanathan and L.Q. Chen, *Coarsening of ordered intermetallic preci-*

pitates with coherency stress, Acta Mater. **50**, 4061-4073 (2002).

52. Y. Ma and A.J. Ardell, *Coarsening of γ (Ni–Al solid solution) precipitates in a γ (Ni3Al) matrix; a striking contrast in behavior from normal γ/γ alloys*, Acta Mater. **52**, 1335-1340 (2005).

53. L.Q. Chen and J. Shen, *Applications of semi-implicit Fourier-spectral method to phase field equations*, Comp. Phys. Comm. **108**, 147-158 (1998).

54. Y. Wang and A.G. Khachaturyan, *Shape instability during precipitate growth in coherent solids*, Acta Metal. Mater. **43**, 1837-1857 (1995).

55. K. Thornton, N. Akaiwa, and P.W.Voorhees, *Large-scale simulations of Ostwald ripening in elastically stressed solids: I. Development of microstructure*, Acta Mater. **52**, 1353-1364 (2004).

56. K. Thornton, N. Akaiwa, and P.W.Voorhees, *Large-scale simulations of Ostwald ripening in elastically stressed solids. II. Coarsening kinetics and particle size distribution*, Acta Mater. **52**, 1365-1378 (2004).

57. J. Boisse, N. Lecoq, R. Patte, and H. Zapolsky, *Phase-field simulation of coarsening of γ precipitates in an ordered γmatrix*, Acta Mater. **55**, 6551-6158 (2007).

58. L. D. Landau and E. M. Lifshitz, *Statistical Physics, Part I* (Butterworth-Heinemann, Oxford, 2000).

Chapter 5

A Renormalization Group Like Model
for a Democratic Dictatorship

Serge Galam*
*CEVIPOF - Centre for Political Research,
Sciences Po and CNRS,
98, rue de l'Université, 75007 Paris, France
serge.galam@sciencespo.fr*

We review a model of sociophysics which deals with democratic voting in bottom up hierarchical systems. The connection to the original physical model and technics are outlined underlining both the similarities and the differences. Emphasis is put on the numerous novel and counterintuitive results obtained with respect to the associated social and political framework. Using this model a real political event was successfully predicted with the victory of the French extreme right party in the 2000 first round of French presidential elections. The perspectives and the challenges to make sociophysics a predictive solid field of science are discussed.

Contents

*serge.galam@sciencespo.fr

1. Introduction

Initiated more than three decades ago,[1–4] sociophysics is today a very active field of research among physicists all over the world with a great deal of papers published in the leading international physical journals. A few reviews and books are now available.[5–7] In this Chapter the focus is on Galam contributions to the field.

Numerous topics are addressed like coalition dynamics among countries,[8] terrorism[9,10] and opinion dynamics.[11–13] It is worth to emphasize that in the earliest times of sociophysics, the physicist community was adamantly opposed to the very principle to mix social sciences with physics.[14]

Here we focus on democratic voting in bottom up hierarchical systems.[2,15] We consider a population made of two parties A and B from which a bottom up hierarchy is built using local majority rules combined with some "natural" inertia bias. Tree like networks are thus constructed, which combine a random selection of agents at the bottom from the surrounding population with an associated deterministic outcome at the top. The scheme relates on adapting real space renormalization group technics to build a social and political structure.

The effective democratic balance of hierarchical organizations based on local bottom up voting using local majority rule is studied. It turns out that several anti-democratic effects are singled out leading to the setting of a democratic dictatorship model.[2,15–22] Counterintuitive aspects of democratic political frameworks are obtained. Using the model a major real political event was successfully predicted with the victory of the French extreme right party in the 2000 first round of French presidential elections.[24,25]

A precise connection to the original physical model is made outlining similarities and differences with an emphasis on eventual novelties with respect to the statistical physics counterparts. To conclude the making of sociophysics as a predictive solid field of science is discussed, emphasizing both the challenges and the risks.

2. Hierarchical bottom-up democratic voting

Consider a population with two competing parties A and B, whose respective proportions are α_0 and $(1 - \alpha_0)$. It could be either a political group, a firm, or a whole society. At this stage each member of the population holds an opinion. A bottom up hierarchy is then built by extracting randomly some agents from the surrounding population. These agents are distributed randomly in a series of groups with finite size r, which constitute the hierarchy bottom. It is the level 0 of the hierarchy. From now on we will use the political case.

Once all groups are formed, each one elects a representative according to a majority voting rule $V_r(\alpha_0)$, which is a function of the current composition of the group. The group compositions being probabilistic with randomly selected members from the surrounding population, they depends on α_0 leading to

$$\alpha_1 = V_r(\alpha_0) \tag{2.1}$$

for the probability to have a A elected and $(1 - \alpha_1)$ for a B as shown in Figure 1 for the case of a group of size 3 composed with 2A and B.

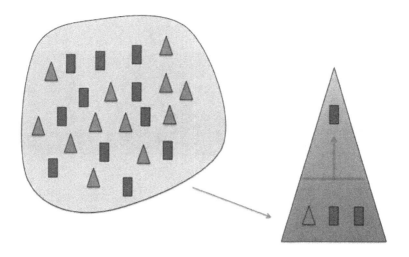

Figure 1. Two parties A and B represented respectively by rectangles and triangles are competing. A one level hierarchy is built with three agents randomly selected from the population. Then, the three chosen agents elect the president using a majority rule. Here two A (rectangle) and one B (triangle) elect a A president.

Figure 2 shows the 4 possible compositions for a group of size 3. The cases with a minority of one against a majority of 2 can come under 3 different configurations in term of order of the agent. However those local orders have no effect on the vote outcome using a majority rule. In this case of 3 agents per group. The associated voting function $V_3(\alpha_0)$ writes,

$$\alpha_1 \equiv V_3(\alpha_0) = \alpha_0^3 + 3\alpha_0^2(1 - \alpha_0) , \tag{2.2}$$

where α_0 is the proportion of elected A persons in the population.

Figure 2.   For a one level hierarchy with three agents randomly selected from a population with A and B agents four different cases are possible as show on the Figure from right to left: $3A \rightarrow A$, $2A$ - $B \rightarrow A$, A - $2B \rightarrow B$, $3B \rightarrow B$.

2.1. *Probabilistic democratic balance*

Figures 1 and 2 show a one-level hierarchy with a probabilistic outcome for the president. However, the pyramid can be extended by adding one hierarchical level, which in turn implies to have 3 groups of 3 agents at the bottom level as shown in Figure 3. There, the ensemble of 3 elected representatives constitute the first level of the hierarchy. They vote to elect the president using also a majority rule. The associated probability is,

$$\alpha_2 \equiv V_3(\alpha_1) = \alpha_1^3 + 3\alpha_1^2(1 - \alpha_1) \ , \qquad (2.3)$$

where $\alpha_1 = V_3(\alpha_0)$.

One illustration is shown in Figure 3 with 3A and 6B at the bottom with a B elected at the top.The bottom majority has won the presidency. In contrast, Figure 4 show a case with 5A and 4B at the bottom with yet a B elected at the top. The bottom minority won the presidency.

Figure 3. A two level hierarchy with groups of 3 persons. It requires 13 agents, 9 at the bottom, 3 elected representative at the first and intermediate level and the president at the bottom. The bottom majority 6B - 3A elects a B (triangle) president. However the outcome is probabilistic as sown in next Figure.

2.2. *From probabilistic to deterministic democratic balance*

Above two cases from Figures 3 and 4 illustrate the probabilistic character of the hierarchical voting. However the probabilistic character can be

Figure 4.   A two level hierarchy with groups of 3 persons. It requires 13 agents, 9 at the bottom, 3 elected representative at the first and intermediate level and the president at the bottom. The bottom minority 4B - 5A elects a B (triangle) president showing the outcome is probabilistic.

reduced and eventually erased by increasing the hierarchy size. Accordingly, the voting process can be iterated as many times as required. Each time a level is added the bottom number of groups must be multiplied by r. At each level, elected representatives form new voting groups to elect higher level representatives according to the same voting rule used to elect them. The process is repeated up forth with $\alpha_n = R_3(\alpha_{n-1})$ till one upper level is constituted by one single group, which elects the hierarchy president.[2,15,17–20,22] After n-levels the probability to have a A president elected at level-$(n + 1)$ writes,

$$\alpha_{n+1} \equiv V_3(\alpha_n) = \alpha_n^3 + 3\alpha_n^2(1 - \alpha_n) \, , \qquad (2.4)$$

where α_n is the proportion of elected A persons at level-n. One 3-level hierarchy is exhibited in Figure 5.

To complete the analysis of the majority rule effect in bottom-up hierar-

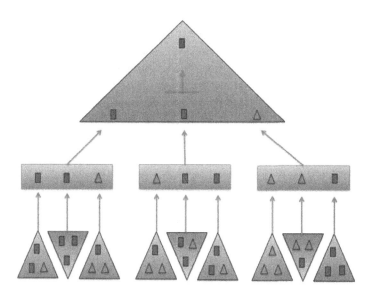

Figure 5. A 3-level hierarchy with 15A and 12B at the bottom and a A president at the top.

chal voting in case of groups of size 3 we focus on the dynamical properties of the voting function $V_3(\alpha_n)$. The associated dynamics is shaped by its fixed points and their respective stabilities. A fixed point is a point α_i which is invariant under the application of the voting function, i. e., $V_3(\alpha_i) = \alpha_i$. A fixed point α_i is stable if $\mid V_3(\alpha_i + \epsilon) - \alpha_i \mid < \epsilon$ while it is unstable when $\mid V_3(\alpha_i + \epsilon) - \alpha_i \mid > \epsilon$ with ϵ being a small real value.

Solving the fixed point equation $V_3(\alpha_i) = \alpha_i$ yields 3 fixed points $\alpha_d = 0$, $\alpha_{c,3} = 1/2$ and $\alpha_t = 1$. The first one corresponds to the disappearance of the A. The last one α_t represents the dictatorship situation where only A are present. Both are stable. In contrast α_c is unstable. It determines the threshold to full power. Starting from $\alpha_0 < 1/2$ repeating voting leads towards (0) while the flow is in direction of (1) for $\alpha_0 > 1/2$.

Therefore majority rule voting produces the self-elimination of any proportion of the A-tendency as long as $\alpha_0 < 1/2$, provided there exists a sufficient number of voting levels. It is therefore essential to determine the number of levels required to ensure full leadership to the initial large tendency.

For instance starting from $\alpha_0 = 0.43$ we get successively $\alpha_1 = 0.40$, $\alpha_2 = 0.35$, $\alpha_3 = 0.28$, $\alpha_4 = 0.20$, $\alpha_5 = 0.10$, $\alpha_6 = 0.03$ down to $\alpha_7 = 0.00$. Therefore 7 levels are sufficient to self-eliminate 43% of the population.

Though the aggregating voting process eliminates a tendency it stays democratic since it is the leading tendency (more than 50%), which eventually gets the total leadership of the organization. The situation is symmetric with respect to A and B. Many countries apply the corresponding winner-takes-all rule, which gives power to the winner of an election.

Figure 6 exhibits two different three agent voting groups hierarchies built from the same population where A has a support of $\alpha_0 = 0.40$ and B a support of 0.60. In the left side hierarchy, six levels restore the democratic balance with $\alpha_1 = 0.00$. It involves 729 agents at the bottom with a total of 1093 agents. The 6-level hierarchy has restored the deterministic outcome of a one vote from the full population with the majority winning. The right side hierarchy has only three levels making the president an A with a probability $\alpha_3 = 0.20$. The bottom involves 27 agents for a total of 40 agents.

The series of Figures 1-6 illustrate the scheme of building a bottom up democratic hierarchies. The first two are the simplest with one level, the presidential one. Third and fourth shows a $n = 2$ hierarchy while the fifth increases the number of levels to $n = 6$. Last one compares a $n = 3$ and a $n = 6$ hierarchies. All of them can be built from the same surrounding population.

2.3. The Status Quo effect

We now move from groups of size 3 to groups of size 4, which embodies the special feature of the possibility to exhibit a tie with 2A-2B configurations for which there exists no majority. The question is then how to solve such a case.

In most social situations it is well admitted that to change a policy required a clear cut majority. In case of no decision, things stay as they are. It is a natural bias in favor of the status quo.[2,15,17-20,22] In real institutions, such a bias is possibly avoided, for instance giving one additional vote to the committee president.

Accordingly the voting function becomes non symmetrical. Assuming the B were in power, for an A to be elected at level $n + 1$ the probability

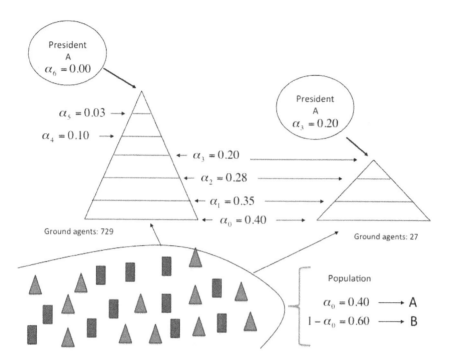

Figure 6.   Two different three agent voting groups hierarchies built from the same population where A has a support of $\alpha_0 = 0.40$ and B a support of 0.60. In the left side hierarchy, six levels restore the democratic balance with $\alpha_1 = 0.00$. It involves 729 agents at the bottom with a total of 1093 agents. The right side hierarchy has only three levels making the president an A with a probability $\alpha_3 = 0.20$. The bottom involves 27 agents for a total of 40 agents.

becomes,

$$\alpha_{n+1} \equiv V_4(\alpha_n) = \alpha_n^4 + 4\alpha_n^3(1 - \alpha_n) , \qquad (2.5)$$

while for B it is,

$$1 - V_4(\alpha_n) = \alpha_n^4 + 4\alpha_n^3(1 - \alpha_n) + 2\alpha_n^2(1 - \alpha_n)^2 , \qquad (2.6)$$

where the last term embodies the bias in favor of B. The associated stable fixed points are unchanged at 0 and 1. However the unstable one is drastically shifted to,

$$\alpha_{c,4} = \frac{1 + \sqrt{13}}{6} \approx 0.77 , \qquad (2.7)$$

which sets the threshold to power for A at a much higher value than the expected fifty percent. In addition, the process of self-elimination is accelerated. For instance starting from $\alpha_0 = 0.69$ yields the series $\alpha_1 = 0.63$, $\alpha_2 = 0.53$, $\alpha_3 = 0.36$, $\alpha_4 = 0.14$, $\alpha_5 = 0.01$, and $\alpha_6 = 0.00$. The series shows how 63% of a population disappears democratically from the leadership levels within only 5 voting levels.

Using an a priori reasonable bias in favor of the B turns a majority rule democratic voting to an effective dictatorship outcome. Indeed to get to power the A must pass over 77% of support, which is almost out of reach in any democratic environment.

3. The killing threshold

For voting groups of any size r the voting function $\alpha_{n+1} = V_r(\alpha_n)$ writes,

$$V_r(\alpha_n) = \sum_{l=r}^{l=m} \frac{r!}{l!(r-l)!} \alpha_n^l (1+\alpha_n)^{r-l} , \qquad (3.1)$$

where $m = (r+1)/2$ for odd r and $m = (r+1)/2$ for even r to account for the B favored bias. The two stable fixed points $\alpha_d = 0$ and $\alpha_t = 1$ are independent of the group size r. The unstable $\alpha_{c,r} = 1/2$ is also independent of the group size r for odd values of r, for which there exist no bias. On the contrary it does vary with r for even values. It starts at $\alpha_{c,2} = 1$ for $r = 2$, decreases to $\alpha_{c,4} = (1+\sqrt{13})/6 \approx 0.77$ for $r = 4$ and then keeps on decreasing asymptotically towards $1/2$ from above.[2,15,17-20,22]

When $\alpha_0 < \alpha_{c,r}$ we can calculate analytically the critical number of levels n_c at which $\alpha_{n_c} = \epsilon$ with ϵ being a very small number. This number determines the level of confidence for the prediction to have no A elected from level n and higher, i.e., only B elected. To make the evaluation we first expand the voting function $\alpha_n = V_r(\alpha_{n-1})$ around the unstable fixed point $\alpha_{c,r}$ with

$$\alpha_n \approx \alpha_{c,r} + (\alpha_{n-1} - \alpha_{c,r})\delta_r , \qquad (3.2)$$

where $\delta_r \equiv dV_r(\alpha_n)/d\alpha_n|_{\alpha_{c,r}}$ with $V_r(\alpha_c) = \alpha_{c,r}$. It can rewritten as

$$\alpha_n - \alpha_{c,r} \approx (\alpha_{n-1} - \alpha_{c,r})\delta_r , \qquad (3.3)$$

which then can be iterated to get

$$\alpha_n - \alpha_{c,r} \approx (\alpha_0 - \alpha_{c,r})\delta_r^n . \qquad (3.4)$$

The critical number of levels n_c at which $\alpha_n = \epsilon$ is then extracted by taking the logarithm on both sides to obtain

$$n_c \approx -\frac{\ln(\alpha_c - \alpha_0)}{\ln \delta_r} + n_0 \, , \tag{3.5}$$

where $n_0 \equiv \ln(\alpha_{c,r} - \epsilon)/\ln \delta_r$. Putting in $n_0 = 1$ while taking the integer part of the expression yields rather good estimates of n_c with respect to the exact estimates obtained by iterations.

Above expression is interesting but does not allow to define a winning strategy from the viewpoint of either A or B since it determines the number of levels required to win the presidency given a bottom support, when indeed every existing organization has a well defined number of levels and that fact cannot be modified at will. Thus, if a party cannot modify the structure of the pyramid, the problem should be defined in a different manner with a reverse perspective. Accordingly, to make the analysis useful in terms of winning strategy the question of

- How many levels are needed to get a tendency self eliminated?

becomes instead

- Given n levels what is the necessary overall support in the surrounding population to get full power with certainty?

Keep in mind that situations for respectively A and B tendencies are not always symmetric. In particular they are not symmetric for even size groups. Here we address the dynamics of voting with respect to the A perspective. To implement the reformulated operative question, we rewrite Eq. (3.4) as,

$$\alpha_0 = \alpha_{c,r} + (\alpha_n - \alpha_{c,r})\delta_r^{-n} \, , \tag{3.6}$$

from which two different critical thresholds are obtained. The first one is the disappearance threshold $\alpha_{d,r}^n$ which gives the value of support under which the A disappears with certainty from elected representatives at level n, which is the president level. In other words, the elected president is a B with certainty. It is obtained putting $\alpha_n = 0$ in Eq. (3.6) with,

$$\alpha_{d,r}^n = \alpha_{c,r}(1 - \delta_r^{-n}) \, . \tag{3.7}$$

In parallel putting $\alpha_n = 1$ again in Eq. (3.6) gives the second threshold $\alpha_{f,r}^n$ above which A gets full and total power at the presidential level. Using

Eq.(11), we get,

$$\alpha_{f,r}^n = \alpha_{d,r}^n + \delta_r^{-n} . \qquad (3.8)$$

The existence of the two thresholds $\alpha_{d,r}$ and $\alpha_{f,r}$ produces a new region $\alpha_{d,r}^n < \alpha_0 < \alpha_{f,r}^n$ in which the A neither disappears totally nor get full power with certainty. There α_n is neither 0 nor 1. It is therefore a region where some democracy principle is prevailing since results of an election process are still probabilistic. No tendency is sure of winning making alternating leadership a reality.

Its extension is given by δ_r^{-n} as from Eq. (3.8). It shows that the probability region shrinks as a power law of the number n of hierarchical levels. Having a small number of levels puts higher the threshold to a total reversal of power but simultaneously lowers the threshold for democratic disappearance.

To get a practical feeling from Eq. (3.8) we look at the case $r = 4$ where we have $\delta = 1.64$ and $\alpha_{c,4} \approx 0.77$. Considering $n = 3, 4, 5, 6, 7$ level organizations, $\alpha_{d,r}^n$ equals to respectively 0.59, 0.66, 0.70, 0.73 and 0.74. In parallel $\alpha_{f,r}^n$ equals 0.82, 0.80, 0.79, 0.78 and 0.78. The associated range extension is 0.23, 0.14, 0.09, 0.05, 0.04. These series emphasizes drastically the dictatorship character of the bottom up voting process.

4. A large scale simulation

To exhibit the phenomena a series of snapshots from a numerical simulation done with Wonczak[21] with 16384 agents are shown in Figures 7, 8, 9, 10. The two A and B parties are represented respectively in white and black squares with the bias in favor of the blacks. The bottom up hierarchy operates with voting groups of size 4 and has 8 levels including the hierarchy bottom.

Four different initial bottom proportions of A and B are shown. On the first three pictures, a huge bottom white square majority is seen to get self-eliminated rather quickly. Written percentages on the lower right part are for the white representation at each level denoted (8) for the bottom and (1) for the president. The "Time" and "Generations" indicators should be discarded.

Figure 7 shows 52.17% bottom people in support of the A (white sites in the Figure), a bit over the expected 50% democratic threshold to take over. However, 3 levels higher no white square appears. The bottom majority has self-evaporated.

Figure 8 shows the same population with now a substantial increase in A (white sites in the Figure) support with a majority of 68.62%, rather more than the democratic balance of 50%. And yet, after 4 levels no more white (A) square is found.

The situation has worsened in Figure 9 where the A (white sites in the Figure) support has climbed up to the huge value of 76.07%. But again, 7 levels higher a B (black) is elected with certainty though its bottom support is as low as (black) 23.03%.

Finally Figure 10 shows an additional very small 0.08% increase in A (white sites in the Figure) support, putting the actual support at 77.05%, which in turn prompts a A to get elected (white site in the Figure) president.

These simulations provide some insight about the often observed blindness of top leaderships towards huge and drastic increase of dissatisfaction at the bottom level of an organization. Indeed it is seen how and why a president, who would get some information about the possible disagreement with its policy, cannot recognize the real current state for support in the population. As seen from Figure, while the opposition is at already a height of 68.62%, the president gets 100% of totally satisfied votes from the two levels below it. Accordingly it will conclude at an overwhelming satisfaction, so why to make any policy change?

5. From 2 to 3 competing parties

Up to now we have treated very simple cases to single out main trends produced by democratic voting aggregating over several levels. In particular we have shown how these thresholds become non symmetric. Such asymmetries are indeed always present in most realistic situations in particular when more than two groups are competing.

Let us consider for instance the case of three competing groups A, B and C.[16] Assuming a 3-cell case, now the (A B C) configuration is unsolved using majority rule as it was for the precedent (A A B B) configuration. For the A B case we made the bias in favor of the group already in power, like giving an additional vote to the committee president.

For multi-group competitions the situation is different. Typically the bias results from parties agreement. For instance, in most cases the two largest parties, say A and B are hostile one to the other while the smallest one C could compromise with either one of them. Then the (A B C)

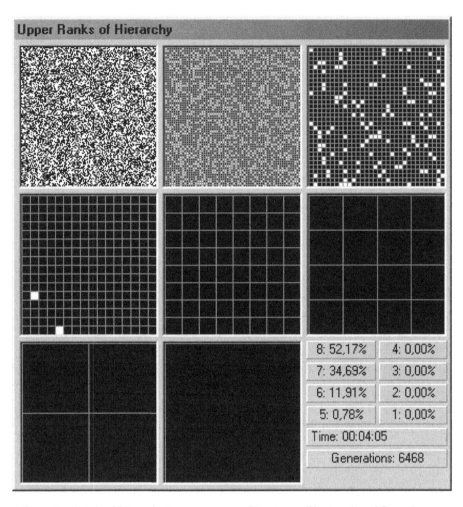

Figure 7.   A 8 level hierarchy for even groups of 4 persons. The two A and B parties are
represented respectively in white and black with the bias in favor of the black squares, i.
e., a tie 2-2 votes for a black square. Written percentages are for the white representation
at each level. The "Time" and "Generations" indicators should be discarded. The initial
white support is 52.17%.

configuration gives a C elected. In such a case, we need 2A or 2B to elect
respectively an A or a B. Otherwise a C is elected. Therefore the elective
function for A and B are the same as for the AB $r = 3$ model. It means
that the critical threshold to full power to A and B is 50%. Accordingly, for
initial A and B supports, which are lower than 50% the C gets full power

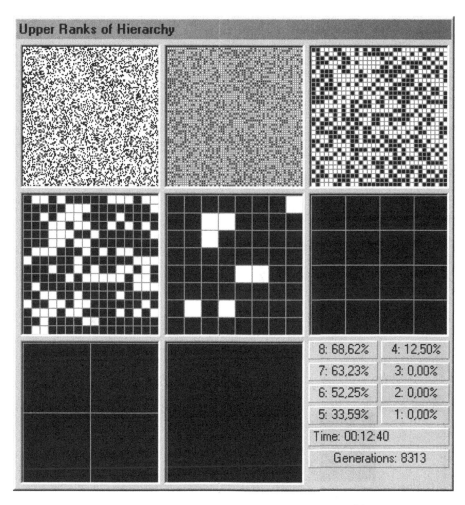

Figure 8.   The same as Figure 6 with an initial white support of 68.62%. The presidency stays black.

provided the number of levels is larger than some minimum limit.[16]

A generalization to as many groups as wanted is possible. However the analysis becomes very quickly much more heavy and must be done numerically. But the mean features of voting flows towards fixed point are preserved.

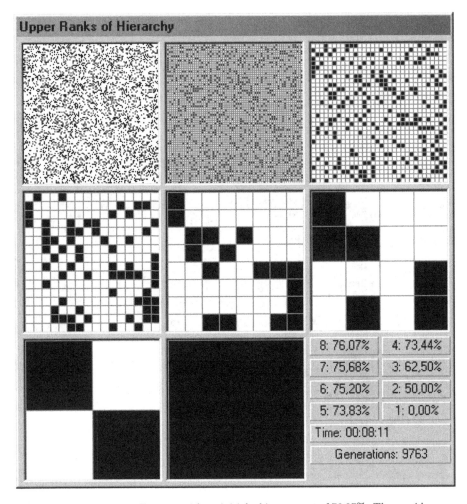

Figure 9.   The same as Figure 6 with an initial white support of 76.07%. The presidency
stays black.

6. The loose connection to physics

The model used does not have a direct statistical physics counterpart. Nevertheless it borrows from it two different features. The first one is to consider a mixture with two species A and B at fixed densities, but such a
situation is not specific to physical models. Moreover it is worth to emphasize that although we are dealing with two species A and B, our agents
are not Ising like variables. Each agent belongs to one party and does not

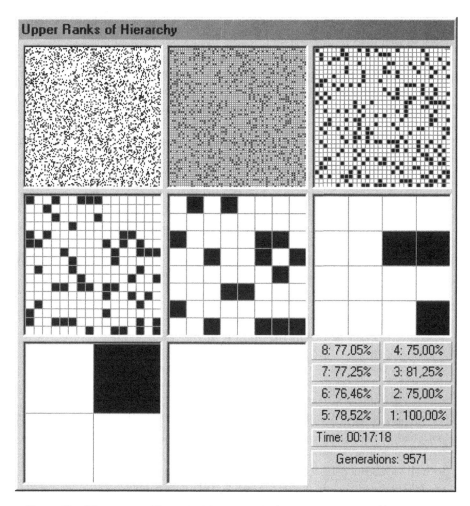

Figure 10. The same as Figure 6 with an initial white support of 77.05%. The presidency finally turned white.

change its affiliation. We are using a mixture of one state variables.

The second borrowed feature is the mathematical local bare mechanism of real space renormalization group scheme, which uses a majority rule to define a super spin. But the analogy ends there since our implementation is performed in a totally different way than in physics.

In our voting case, the local rule is operated to add a real new agent above the given voting group. This agent does not substitutes to the group. It is not a virtual super spin. The group and its elected representative are real agents, which are simultaneously present. All the hierarchy levels are real, in the sense of the model.

In physics renormalization group technics are a mathematical method.[23] It is used to extract rather accurately some physical quantities, which characterize a given system. Applying a renormalization group scheme to a physical system does not modify the system. In our case the system is built step by step following the scheme. The hierarchy does exist, it is the system. Accordingly a n level hierarchy is different from a m level hierarchy.

Given some proportions of respectively A and B agents in a population we build a bottom up voting hierarchy by selecting agents from that population. At the bottom level of the hierarchy agents are randomly selected from the surrounding population to constitute the voting groups. Afterwards at the first level, the elected representatives are selected from the population according to the vote outcomes, i.e., the party affiliation is imposed by the voting of the group beneath. They are not randomly selected.

7. Unexpected political implementation

Although the model is only a snapshot of real hierarchies it grasps some essential and surprising mechanisms of majority rule voting. It is very generic and allows to consider many different applications.

In particular it exhibits several counterintuitive results and provides paradoxical and unexpected explanations to a series of social features and historical events. Especially the empirical difficulty in changing leaderships in well established institutions. It also allows to shed a new light on an astonishing and crucial historical event of the last century, the sudden and quick auto-collapse of eastern European communist parties.

Up to this historical and drastic end, communist parties in power has seemed to be eternal. Once they collapsed all once many explanations were given to base the phenomena on some hierarchical opportunistic change within the various organizations. Among others one reason for the eastern European countries was the end of the Soviet army threat.

However our hierarchical model may provide some different new insight at such a unique event. Communist organizations are indeed based, at least in principle, on the concept of democratic centralism which is a tree-like hierarchy similar to our bottom up model. Suppose for instance that the critical threshold to power was of the order of 77% like in our size 4 case. We could then consider that the internal opposition to the orthodox leadership did grow continuously over several decades to eventually turn massive, yet without any visible change at the top organizations. Then at some point of internal opposition, its internal increase has reached the critical threshold. There a little more increase would at once produce a surprising shift of the top leadership as exhibited in our series of Figures 7-10. From outside, the decades long increase of opposition was invisible. Indeed it looks like nothing was changing. And once the threshold passed, the shift appears as instantaneous.

Therefore, what looked like a sudden and punctual decision of a top leadership could be indeed the result of a very long and solid phenomenon inside the communist parties. Such an explanation does not oppose to the very many additional features which were instrumental in these collapses. It only singles out some trend within the internal mechanism of these organizations, which in turn made them extremely stable.

Using our model we predicted a political scenario, which could happen in France, and which eventually did occur with respect to the extreme right party National Front. We enumerate the conditions for its success in 1997[24,25] and it happened along these line in 2000 with its leader winning the presidential first round. He eventually lost in the second final run.

8. Conclusion

Using our simple but not trivial model of bottom-up democratic voting comment we obtained an unexpected explanation to the last century auto-collapse of eastern communist parties. Contrary to explanations related to a combination of an opportunistic change within the various organizations with the disappearance of the Soviet army threat, your conclusion shows an alternative enlightening a long and slow process.

At this stage it is of importance to stress that modeling social and political phenomena is not aimed to state an absolute truth but instead to single out some basic trends within a very complex situation. All models can be put at stake, and changed eventually. The serious challenge of sociophysics is to prove itself it can become a predictive science with well established elementary rules of social and political behaviors. The task is hard but rather exciting.

References

1. S. Galam, Y. Gefen and Y. Shapir, *Sociophysics: A mean behavior model for the process of strike*, Math. Journ. of Sociology **9**, 1-13 (1982).
2. S. Galam, *Majority rule, hierarchical structures and democratic totalitarism: a statistical approach*, Journ. of Math. Psychology **30**, 426-434 (1986).
3. S. Galam and S. Moscovici, *Towards a theory of collective phenomena: Consensus and attitude changes in groups*, Eur. Journ. of Social Psychology **21**, 49-74 (1991).
4. S. Galam, *Rational group decision making: a random field Ising model at $T = 0$*, Physica A, **238**, 66-80 (1997).
5. C. Castellano, S. Fortunato and V. Loreto, *Statistical physics of social dynamic*, Reviews of Modern Physics, **81** 591-646 (2009).
6. B.K. Chakrabarti, A. Chakraborti, A. Chatterjee (Eds.), *Econophysics and Sociophysics: Trends and Perspectives*, Wiley-VCH Verlag (2006).
7. S. Galam, *Sociophysics: a physicist's modeling of psycho-political phenomena*, Springer (2012).
8. S. Galam, *Fragmentation versus stability in bimodal coalitions*, Physica A **230**, 174-188 (1996).
9. S. Galam, *The September 11 attack: A percolation of individual passive support*, Eur. Phys. Journ. B **26** Rapid Note, 269-272 (2002).
10. S. Galam and A. Mauger, *On reducing terrorism power: a hint from physics*, Physica A **323**, 695-704 (2003).
11. S. Galam, *Public debates driven by incomplete scientific data: The cases of evolution theory, global warming and H1N1 pandemic influenza*, Physica A **389**, 3619-3631 (2010).
12. S. Galam, *Collective beliefs versus individual inflexibility: The unavoidable biases of a public debate*, Physica A **390**, 3036-3054 (2011).
13. S. Galam, *Modeling the Forming of Public Opinion: an approach from Sociophysics*, Global Economics and Management Review **18**, 2-11 (2013).
14. S. Galam, *Sociophysics: a personal testimony*, Physica A **336**, 49-55 (2004).
15. S. Galam, *Social paradoxes of majority rule voting and renormalization group*, Journ. Stat. Phys. **61**, 943-951 (1990).
16. S. Galam, *Political paradoxes of majority rule voting and hierarchical systems*, Int. Journ. General Systems **18**, 191-200 (1991).
17. S. Galam, *Application of Statistical Physics to Politics*, Physica A **274**, 132-139 (1999).
18. S. Galam, *Real space renormalization group and totalitarian paradox of majority rule voting*, Physica A **285**, 66-76 (2000).
19. S. Galam and S. Wonczak, *Dictatorship from Majority Rule Voting*, Eur. Phys. Journ. B **18**, 183-186 (2000).
20. S. Galam, *How to Become a Dictator*, in *Scaling and disordered systems. International Workshop and Collection of Articles Honoring Professor Antonio Coniglio on the Occasion of his 60th Birthday*. F. Family. M. Daoud. H.J. Herrmann and H.E. Stanley, Eds., World Scientific, 243-249 (2002).
21. S. Galam, *Stability of leadership in bottom-up hierarchical organizations*,

Journ. of Social Complexity **2** 62-75 (2006).

22. S. Galam, *The Drastic Outcomes from Voting Alliances in Three-Party Democratic Voting (*1990 → 2013*)*, Journ. Stat. Phys. **151**, 46-68 (2013).

23. B. Hu, *Introduction to real-space renormalization-group methods in critical and chaotic phenomena*, Physics Reports **91**, 233-295 (1982).

24. S. Galam, *Le dangereux seuil critique du FN*, £e Monde, Vendredi 30 Mai, 17 (1997).

25. S. Galam, *Crier, mais pourquoi*, Libération, Vendredi 17 Avril, 6 (1998).

Subject Index

Printed in the United States
By Bookmasters